Geotechnics of Landfills and Contaminated Land

Ernst & Sohn

Geotechnics of Landfills and Contaminated Land
Technical Recommendations "GLC"

Edited by the German Geotechnical Society
for the International Society of Soil Mechanics
and Foundation Engineering

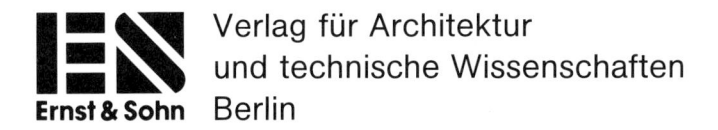

Ernst & Sohn — Verlag für Architektur und technische Wissenschaften Berlin

Die Deutsche Bibliothek – CIP-Einheitsaufnahme

Geotechnics of Landfills and Contaminated Land:
Technical Recommendations "GLC" / ed. by
the German Geotechnical Society for the International
Society of Soil Mechanics and Foundation Engineering. –
Berlin: Ernst 1991
       ISBN 3-433-01214-8
NE: Deutsche Gesellschaft für Erd- und Grundbau

© 1991 Ernst & Sohn Verlag für Architektur und technische Wissenschaften, Berlin.

All rights reserved, especially those of translation into other languages. No part of this book shall be reproduced in any form – i. e. by photocopying, microphotography or any other process – or be rendered or translated into a language useable by machines, especially data processing machines, without the written permission of the publishers.

The quotation of trade descriptions, trade names or other symbols in this book shall not justify the assumption that they may be freely used by anyone. On the contrary, these may be registered trademarks or other registered symbols, even if they are not expressly marked as such.

Composition and Printing: Oskar Zach GmbH & Co. KG, D-1000 Berlin 31
Binding: Lüderitz & Bauer GmbH, D-1000 Berlin 61

Printed in the Federal Republic of Germany

Preface

Although there is a strong demand to avoid, to reduce and to recycle waste, landfills will be indispensable within any sound waste disposal concept. It is an important task of environmental geotechnics to establish principles in the design and construction of landfills, in particular with respect to long term safety. Further, geotechnics are strongly involved in remedial works for abandoned landfills and contaminated land due to former industrial activities.

It is an important challenge for geotechnical engineers and scientists to contribute within their profession to the protection of air, soil and water. New pollutions and contaminations have to be avoided, as well as existing contaminations in the soil have to be cleaned up. For these new and very complex problems appropriate solutions should be prepared by the geotechnical community, especially by geotechnical societies.

In 1985 the German Geotechnical Society (Deutsche Gesellschaft für Erd- und Grundbau e. V., Essen) established the working group AK 11 for Geotechnics of Landfills and Contaminated Land (Geotechnik der Deponien und Altlasten). The 16 members (I)*) of this committee are experts in environmental geotechnics and represent construction industry, universities, consulting firms and administrations. The results of the committee work are fixed in 20 recommendations (GDA-Empfehlungen) which were published in 1990.[1])

At the International Conference on Soil Mechanics and Foundation Engineering (ICSMFE 1989) in Rio de Janeiro, Professor Smoltczyk, Vice President for Europe, founded a European Technical Committee to work on geotechnics within environmental protection activities, in particular on Geotechnics of Landfills and Contaminated Land. The European Technical Committee No. 8 (ETC 8) consists of 6 members of several European countries (II)*). The chairman is presented by the sponsoring German Geotechnical Society.

ETC 8 started with its first meeting in December 1989. The main objective of this committee is to provide the engineering practice with recommendations for its work. It is planned to present the first report of ETC 8 at the European Conference on Soil Mechanics and Foundation Engineering in May 1991 in Florence.

During the first meeting it was decided to base the committee work on the recommendations of the working group AK 11 of the German Geotechnical Society. A useful help was the draft translation of these recommendations which had been

---

*) The names are listed on page VII

[1]) GDA – Empfehlungen des Arbeitskreises „Geotechnik der Deponien und Altlasten" der Deutschen Gesellschaft für Erd- und Grundbau. Verlag Ernst & Sohn, Berlin 1990

prepared by the British Ministery of Environmental Protection. These German recommendations were improved taking into consideration the European challenge after 1992. In summer 1990, a one-week workshop was held in Bochum, Germany, with additional experts as guests (III).*)

The result of this workshop is this book containing the draft version of the GLC-recommendations on *G*eotechnics of *L*andfills and *C*ontaminated Land. These recommendations should be a European-wide framework for environmental geotechnical activities in spite of different legal conditions in the European countries. Specific regulations and standards of a number of countries are mentioned in the references which are acknowledged in the text by letters [A to CC]; references with numbers (1 to 11) are related to literature for special problems.

The book contains 18 recommendations and an appendix; the appendix was worked out on the basis of a proposal of the Italian group. Each recommendation can be related to one of the following subjects:

R1 Recommendations on the Site Assessment
R2 Recommendations on the Principles of Design
R3 Recommendations on Suitability Testing
R5 Recommendations on Quality Assurance.

The R4 recommendations, which will be related to the construction and to the production activities on site, are under preparation.

This book should serve as a basis for discussion and mutual agreement. The members of the geotechnical community are encouraged to send their comments, questions and corrections not later than October 1st, 1991 to the following address:

Prof. Dr.-Ing. Hans L. Jessberger
Ruhr-Universität Bochum
Postbox 10 21 48
4630 Bochum, Germany

The excellent cooperation of the ETC 8-members, the assistance of the participants to the workshop and of all other colleagues who gave us valuable advice are gratefully acknowledged. Many thanks also to the publishing company which made an excellent job.

We hope that these recommendations will be successfully used in Europe and abroad.

Hans L. Jessberger
Chairman ETC 8

---

*) The names are listed on page VII

List of Names

I.   AK 11 – Members 1989

Hans L. Jessberger (Chairman)
Erwin Gartung
Wolfgang Hollstegge
Ewald E. Kohler
Klaus Krubasik
Quirin Laumans
Holger Meseck
Hermann K. Neff
Manfred Nußbaumer
Manfred Reinhardt
Volker Schenk
Heinz Steffen
Siegfried Steinkamp
Klaus Stief
Dieter Stroh
Karl R. Ulrichs

II. ETC 8 – Members 1990

CH– Christian Schlüchter
F – Bertrand Soyez
G – Hans L. Jessberger (Chairman)
I – Mario Manassero
NL– Ton Puthaar
UK– Andrew Street

III. ETC 8 – Guest experts 1990

F – Jean-Louis Bostvironnois
G – Ewald E. Kohler, Holger Meseck, Hermann K. Neff
I – Francesco Belfiore
NL– Ton Jacobs
UK– Keith J. Seymour
     Ralph Kockel (Secretary)

Content

Preface V

List of Names VII

R 1 Recommendations on the Site Assessment 1
    R 1-1 Type and Scope of Geotechnical Testing for Site Investigation 1
    R 1-2 Groundwater, Soil and Air Sampling and Sample Treatment for the Survey of Waste Disposal Sites and Contaminated Sites ... 7

R 2 Recommendations on the Principles of Design 12
    R 2-1 Geotechnical Principles in the Design of Landfills 12
    R 2-2 Geotechnical Principles in the Design of Remedial Action 14
    R 2-3 Composite Basal Lining System 14
    R 2-4 Capping System 17
    R 2-5 Health and Safety Considerations at the Contract Stage for Rehabilitation of Contaminated Sites 20
    R 2-6 Waste Body Stability 22

R 3 Recommendations on Suitability Testing 25
    R 3-1 Suitability Testing of Mineral Capping and Basal Seals 25
    R 3-2 Suitability Testing for Mineral Sealing Compounds 29
    R 3-3 Clay Mineralogical Characterisation of the Mineral Basal Seals . 34
    R 3-4 Chemical Impact of Leachate on Mineral Sealing Materials 36
    R 3-5 Field Trials for Mineral Basal Seals and Caps 37
    R 3-6 Suitability Testing of Waste Substances for Placement in the External Stability Zone of Landfills 41

R 5 Recommendations on Quality Assurance 44
    R 5-1 Quality Assurance Principles 44
    R 5-2 Quality Assurance of Subgrade, Mineral Capping and Basal Sealing Layers 45

R 5-3 Quality Assurance for Vertical Cut-off Walls 48

R 5-4 Quality Assurance for Placement of Waste in External Stability Zones of Landfills 54

Appendix   **Geomembranes for Composite Cut-off Diaphragm Walls** 56

References 59

Index 75

R 1 Recommendations on the Site Assessment

R 1-1 Type and Scope of Geotechnical Testing for Site Investigation

1 Tasks

1.1 General

Waste disposal sites must be designed and managed in such a way that no harmful substances reach the biosphere and hydrosphere in unacceptable quantities (in accordance with appropriate national regulations). The design concept for a disposal site also depends on the structure and behaviour of the subsoil. Particular attention must be paid to the protection of groundwater and the mechanical stability of the waste material.

The planning, development and management of disposal sites must be suited to the geological and hydrogeological conditions. For proposed new sites, a detailed geological, hydrogeological and geotechnical investigation is therefore essential [A].

The type and size of the site investigation depends on the following factors:
- topography and structure of the area;
- type and behaviour of the waste;
- geological/hydrogeological setting.

At the same time the following must be considered:
- design requirements (see R 2-1);
- overall safety plan (see R 2-1, section 2).

In order to assess the suitability of a disposal site it is essential to have accurate knowledge of the distribution of groundwater flow paths and barriers (aquifers and aquicludes), their hydraulic properties, the deformation behaviour of the subsoil and the potential for improving the sealing effect of the subsoil. Consideration must also be given to the need for setting up adequate controls and undertaking subsequent remedial works if appropriate. To assess and evaluate the behaviour of the subsoil as a foundation for a disposal site (base area or perimeter boundaries) it is essential to have knowledge of the local general geological setting, including the following principal aspects:

- characteristics of the morphology;
- structure, extent and geological age of the outcropping strata;
- tectonic structures;
- deeper subsoil if it comprises cavities or soluble rocks;
- aquifers and groundwater flow;
- risk of earthquakes and other natural hazards.

1.2 Composition and Distribution of Superficial Deposits

In order to assess the subsoil for a disposal site it is necessary to know:
- composition, physical and chemical properties and sequence of strata;
- lateral and vertical continuity and distribution of the strata (facies changes);
- porosity;
- permeability (to water and leachate);
- resistance to erosion and washing away of fine particles;
- stress deformation behaviour.

1.3 Structure and Sequence of Solid Strata

Due to regional geological factors and morphological characteristics, superficial deposits are often relatively thin and therefore the underlying solid strata may have to be included in the survey. Here the following factors need to be considered:
- type of rock, mineralogical composition and stratigraphy;
- state of weathering and weathering resistance;
- solubility in water and leachate or other aggressive solutions;
- type and position of geological boundaries;
- extent, degree of separation and widths of individual joints;
- tectonic and petrographical anisotropies in the rock mass;
- karstification and risk of subsidence;
- deformation behaviour of the rock mass;
- permeability to water, leachate, gases and other aggressive solutions (hydrocarbons, etc).

1.4 Determination of Hydrogeological Data

Disposal sites must be prevented from having unacceptable impacts on groundwater, surface water, and particularly water abstraction sources. Comprehensive knowledge of the groundwater regime is therefore required, including the following detailed information:
- groundwater regime, direction of flow, gradient and rate of flow, including long-term and seasonal fluctuations;
- permeability (horizontal and vertical) or transmissivity of the outcropping strata, with maximum and minimum values;
- distribution, thickness and depth of aquifers, aquicludes and aquitards, including the locations of any springs;
- groundwater levels, indicating hydraulic gradients and effective flow velocity in the individual strata components if appropriate;
- groundwater chemistry, including determination of naturally occurring aggressive substances and groundwater quality;
- groundwater protection zones;
- groundwater abstraction and its effects;
- groundwater abstraction rights;

R 1-1   Geotechnical Testing   3

- influence of short-term or long-term lowering of the water table, restoration and extraction or augmentation of groundwater in the future;
- influence of nearby open waters and their relationship with the groundwater system;
- situation in respect to receiving streams, influence of flooding and tides, if appropriate;
- effective rainfall, surface runoff, percolation rate, evaporation and groundwater recharge.

1.5   Consideration of Special Factors

Artificial interference with the subsoil may have significantly altered the natural conditions. The existence of natural deposits worthy of protection or archaeological factors, may preclude use of the site as a landfill. The following points should be included in the survey:

- the stability of existing slopes if trenches are used;
- possibility of sealing off any old tunnels or adits;
- potential for subsidence caused by abandoned or existing mine workings and/or gas/groundwater extraction wells (underground and surface workings);
- presence of workable natural materials in the subsoil;
- presence of geological features or archaeological monuments worthy of protection;
- where spoil heaps or landfills are used, survey of the underlying exposed natural subsoil.
- background contamination of the subsoil and/or groundwater.

2   Desk Study

The process outlined below is usually adequate for the preparation of a desk study aimed at assessing the general suitability of the site. Under certain circumstances, additional investigations, soundings and/or borings may be necessary.

The desk study should include compilation of all available information from archives, geological and topographical maps, meteorological data, aerial or space photographs (black and white, colour, infra-red). The configuration and previous use of the land, data relating to water supply and distribution and analysis of available borehole data should also be reviewed. In addition to geological and hydrogeological maps, pedological atlases and maps of mineral deposits can also yield valuable information on the subsoil, as can regional geological publications.

It is essential to inspect the terrain, recording and assessing information which relates to the subsoil.

3 Field Surveys For Approval and Construction Planning

3.1 General

The geotechnical survey for a proposed landfill must commence with a careful desk study leading to the production of a programme for field investigation and laboratory testing. The final scope of the overall programme will often not be decided until field surveys are in progress.

Boreholes and trial pits will give direct information on the subsoil, as will shafts and survey tunnels in special cases. Geophysical procedures can be useful as an indirect method of investigation, especially where large areas are to be surveyed. Boreholes provide information on subsoil characteristics at specific locations. Anisotropies, which may occur as a result of changing depositional conditions, require a dense network of boreholes or fairly long trenches in order to assess natural low permeability strata.

In carrying out site investigation work (including sampling and description of soils and rocks) the appropriate national standard should be followed [B]. In addition, a photographic record should be made of all trial pits and samples taken.

The site investigation should cover the whole landfill area and the surrounding area, if appropriate, and should take account of any possible influence on the groundwater. The depth of investigation should also take account of the influence of settlement caused by the waste and the effect of any vertical hydraulic continuity between groundwater systems within the underlying strata.

Where the subsoil has not been previously investigated, an appropriate number of boreholes should be installed (at least one borehole per hectare of landfill is recommended) for approval or planning purposes. If necessary, the perimeter of the landfill and hydrogeology of the surrounding area should also be investigated. These minimum requirements should only be reduced in special circumstances.

The investigation programme should be prepared, supervised and interpreted in a responsible manner by a qualified geotechnical expert. The following investigation and observation techniques may be used.

3.2 Superficial Deposits [C]

- boreholes with continuous sampling;
- exploratory boreholes;
- cone penetrometer testing;
- trial pitting;
- micro-seismology to determine strata thicknesses and general identification of the boundary between solid rock and superficial deposits;
- surface and borehole geophysics for the determination of strata distribution and thickness and assessment of permeability, if appropriate.

3.3 Solid Rock [D]

- cored boreholes, with directional orientation, if appropriate;
- trial pitting;
- water pressure tests;
- optical probe and closed circuit television camera (CCTV);
- borehole geophysics and geo-electrical profile measurements for the general assessment of the structure and permeability of the rock mass (correlated by reference to exploratory boreholes);
- shafts and tunnels in special cases.

3.4 Groundwater [E]

Boreholes can be used for monitoring groundwater levels and quality within the vicinity of the site. The location and depth of piezometers should be chosen such that different groundwater levels or discrete aquifers can be clearly defined. The measurement of groundwater levels, together with groundwater chemistry where appropriate, provides the necessary data for defining a groundwater model.

The measurement of groundwater levels should be undertaken at sufficiently frequent intervals so that fluctuations may be identified and evaluated; the potential effect of the landfill on the groundwater regime should also be considered.

Piezometers should be installed in such a way that groundwater samples can be taken. The diameter selected for the borehole should be sufficient for this purpose (a diameter of 150mm is recommended). Care must be taken to ensure that the piezometer tube is adequately sealed at the surface and between the measurement horizons to prevent water percolating in.

Field tests may be necessary in some case to confirm flow conditions; these may include:

- groundwater tracing tests;
- trial pumping;
- infiltration tests;
- flow measurement.

4 Laboratory Tests

4.1 Petrographical and Soils Testing

Laboratory tests on samples from boreholes or trial pits serve to classify the soil and rock material and to determine stress deformation behaviour and permeability. Laboratory tests must be conducted on an appropriate scale for all relevant soil strata in order to determine variations in behaviour. The testing should be laid down by a qualified geotechnical expert. The testing procedure should be documented to ensure results are recorded and reliable.

Depending on the particular soil type, tests are to be carried out in accordance with Recommendation R 3-1.

In the case of solid rock additional testing may be appropriate, and include:
- particle size distribution, swelling capacity and water intake of the interface;
- permeability of porous rock;
- state of weathering [F];
- solubility and resistance to weathering

4.2 Geochemical Testing

Geochemical testing is used to assess the quality of the relevant groundwater limits and the composition of the strata. The scope of testing should be laid down by the qualified expert taking account of relevant national regulations and in agreement with an experienced chemical investigation laboratory.

5 Presentation of Investigation Results

Site investigation results should be presented in accordance with relevant standards and regulations. With regard to the presentation of results from field investigations, particular reference is made to [G]. In addition for field exploration, the corresponding state of weathering should be indicated [F].

Presentation of the investigation results in the form of graphs or diagrams is recommended, particularly where there is a large volume of data. The following may be considered:
- site plans indicating:
  - location of boreholes, trial pits etc;
  - geological and groundwater level/contour plots;
  - groundwater flow direction and effective flow velocity;
  - groundwater abstraction (including water resource catchment areas and water protection areas);
  - surface water and other hydrological features;
  - geochemical zones for groundwater and soil/rock.
- geological sections (indicating borehole records used);
- spatial profile (geological overlays and block models);
- representation of the groundwater system (rainfall distribution, fluctuations in groundwater level, flood and tidal influence).

In addition to the presentation required by the standards, laboratory test data should be set out in summary tables, for example in accordance with [H]. The presentation of geophysical measurement data should be agreed with the investigation company.

6 Overall Assessment

The results of the site investigation should be subject to an overall analysis and evaluation, taking account of the particular design stage and specific requirements of the general safety plan. This assessment should be set out in a geotechnical report.

R 1-1 Geotechnical Testing

This report must address the following aspects as a minimum:
- description and representation of the geological structure;
- presence and suitability of natural low permeability strata (thickness, depth, horizontal continuity, permeability, adsorption capacity);
- groundwater regime and permeabilities within the area to be landfilled and its environs; a groundwater model may be appropriate;
- stability of natural or artificial slopes;
- bearing capacity and deformation behaviour of the subsoil;
- faults, possible subsidence, risk of collapse, earthquake risk and other hazard situations;
- overall evaluation of the subsoil as a natural barrier for the site;
- notes on geotechnical measures required to improve the properties of the subsoil as a natural safety barrier.

R 1-2 Groundwater, Soil and Air Sampling and Sample Treatment for the Survey of Waste Disposal Sites and Contaminated Sites

1 General

Monitoring locations and sampling methods must be selected in such a way that the site investigations can be carried out in accordance with R 1-1. In addition, it is necessary to ensure that for a contaminated site or an abandoned landfill the initial site conditions or range of aggressive substances present can be reliably determined, effects on the environment can be recognized and remedial measures introduced in good time.

For contaminated sites or abandoned landfills, sampling and sample treatment, together with any necessary safety provisions, should be agreed with an experienced chemical testing laboratory, and related to the aggressive substances expected or found. R 2-5 applies accordingly.

2 Water Sampling

Sampling equipment should be selected such that it has a minimal effect on the quality of the water samples. Brass, chrome-plated, nickel-plated or galvanized material should not be used if water is being tested for heavy metals. When testing for iron and manganese, iron or steel alloys are acceptable. Samples from dye tests must be kept cool in dark bottles prior to analysis.

Particular care is required when sampling volatile organic substances and highly toxic constituents in very low concentrations. This involves the use of special procedures and equipment.

2.1 Monitoring Borehole Construction

Monitoring borehole construction should be carried out in accordance with appropriate national standards. The location of the sampling/monitoring zones should be selected and the installation supervised by a qualified geotechnical engineer or hydrogeologist, based on field data obtained during borehole construction.

In multi-layered aquifers it is recommended that individual monitoring boreholes should be constructed to sample each aquifer unit. This also applies to multi-level sampling within one aquifer.

With care, it is possible to install multiple piezometers, or single piezometers divided into sections by stationery packers to provide samples relating to specific horizons in a single monitoring borehole. However, such installations are prone to vertical leakage problems.

Multiple packer systems are now being developed and tested in which horizontal flow conditions can be achieved by controlling the pumping rate in individual packer zones.

In the case of continuously screened boreholes, samples specific to a particular horizon should not be taken by moveable packers because of the inevitable circulation through the annular space behind the screen.

2.2 Sampling Equipment and Pumps

2.2.1 Bailing Devices

Simple bailing tubes are suitable for rapidly obtaining samples without a high degree of accuracy and for relatively high ionic concentrations.

Bailing devices should not be used in monitoring boreholes where a high degree of accuracy is required because of the risk of disturbance of the water column and the uncertainty of the specific sampling horizons. Other sampling devices are available for use in static water columns. These include mechanical and electro-magnetically operated latch-samplers, which have end-ports which can be sealed once the equipment has been lowered to the selected sampling depth.

2.2.2 Pumps

When obtaining pumped samples it is necessary to maintain a continuous discharge. It is recommended that pumping should continue until the pH value, the electrical conductivity and/or the temperature have attained constant values. Suction pumps, plunger pumps, submersible electrical pumps and gas driven diaphragm pumps can be used. Table 1-2.1 indicates the range of applicability of the various pumping devices.

R 1-2 Sampling

Table 1-2.1
Application of water sampling pumps

Up to 3 m		3–9 m	Deeper than 9 m	Depth of water table
30–50 mm	50–100 mm	Over 100 mm	45–100 mm	Casing diameter
Suction pumps Without | with foot operated valve		Submersible pumps	Reciprocating plunger pumps	Sampling device

a) Suction Pumps

Suction pumps can produce an output of around 1 to 2 l/s but are limited to situations where the head lift relative to ground level is small.
Suction pumps are not suitable for taking samples for dissolved gas determination because of the high degree of induced aeration.

b) Submersible Pumps

The monitoring borehole construction is selected according to R 1-1 such that electrical submersible pumps with a diameter of up to 100 mm approx. can be used.
Petrol driven pumps should be avoided when sampling for low concentrations of hydrocarbons. Low output, small diameter submersible pumps are available which are operated by a 23 volt D.C. battery.

2.3 Further Recommendations for Sampling and Sample Treatment

- Sampling points should be located down-hydraulic gradient from the pollution source and also at least one up-gradient of the site.
- Sampling points should also be located outside the known area of contaminated groundwater, to provide background readings.
- Appropriate adjustment to the vertical sampling interval in thin multi-layered or highly fissured aquifers and where relatively high flow velocities are involved.
- A sample volume of at least 1 litre should be taken, following removal of at least 3 pore – volumes of the gravel pack.

The following points should be borne in mind when handling the samples:
- safety precautions;
- storage of the samples in cool, frost-free conditions;
- for major ion analyses, glass containers should generally be used;
- disposal of water from flushing the piezometers should be in accordance with relevant anti-pollution regulations.

3 Soil Sampling [I]

Appropriate drilling procedures, sampling techniques, sample handling and documentation should be adopted.

When investigating contaminated land or abandoned landfill sites, dry drilling should be carried out if possible. The sampling equipment must be carefully cleaned immediately after use in accordance with the relevant safety provisions.

Soil samples are obtained during the site investigation for proposed landfills in accordance with R 1-1 or during the examination of contaminated land or abandoned landfill sites. Various sampling devices are used in conjunction with the drilling procedures outlined in R 1-1. The choice of device depends on the soil and groundwater conditions and the proposed testing schedule. For transport and storage of non-contaminated samples see Reference [I].

Samples from contaminated land or abandoned landfill sites must be tightly sealed immediately they are taken, or placed in containers which can be sealed in such a way as to minimise the volume of air remaining in the container. The samples should be kept dark and cool and delivered to the chemical testing laboratory without delay.

4 Air Sampling

Air samples, principally from the unsaturated soil zone, serve primarily to locate and define contamination sources and assist in confirming whether groundwater is contaminated.

The *Kanitz-Selenka* procedure [1] is based on extracting the soil pore gas ("soil air") from the soil. Hydrocarbons are passed through a sorption column and they are enriched. For gas chromatographic analysis, adsorbed halogenated volatile hydrocarbons and highly volatile or semi-volatile aromatic and aliphatic hydrocarbons are desorbed using a solvent. Gases, such as CH_4, CO_2, H_2S produced by landfills are either measured in the field with the aid of field measuring equipment or after gas chromatographic separation using a thermal conductivity detector or a flame ionisation detector. The sorption column is set up in the above-ground section of a soil lance as shown in Fig. 1-2.1. The lance is inserted into the soil in a pre-formed hole made by a percussion probe or percussion channel probe. The soil lance is about 1.5 m long, made of high grade steel and has an external diameter of 22 mm. The pump shown in Fig. 1-2.1 permits the removal of soil air by suction at a rate of about 2 l/m. Practical experience has shown that installing a HDPE soil air gauge with a screw cap, instead of the soil lance, is preferred when taking soil air samples over longer periods of time.

R 1-2 Sampling

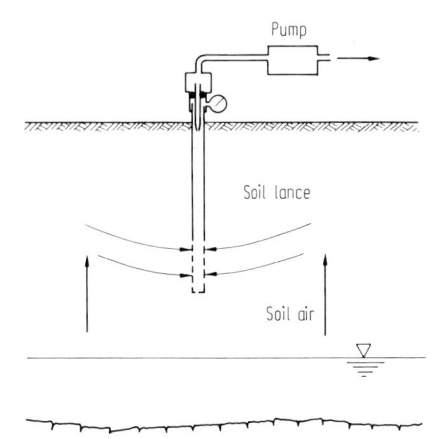

Fig. 1-2.1
Soil air sampling in accordance with [1].

To identify a contaminated groundwater plume, it has proved useful to establish monitoring points approx. 20 to 50 m apart on several axes normal to the direction of groundwater flow. The measurement axes should be approx. 50 to 200 m apart, so that the whole of the contaminated groundwater area can be covered.

Another sampling procedure, which also makes use of a perforated soil lance, is the "gas monitor method" [2]. This method can only be used to depths of up to 3 m. The technique is most suited for an initial general survey and provides a basis for a more detailed investigation programme. It only allows slightly volatile organic substances to be detected.

The interval required between monitoring stations depends on the assumed distance from the contamination source and the groundwater. A sampling regime should be selected in which the monitoring points are spaced approx. 10 m and 50 m apart. In the case of sandy and gravelly soils with no low permeability intercalations, monitoring results relating to groundwater depths of approx. 10 to 15 m are possible from sampling depths of 2 to 3 m.

The soil air sampling probe can be inserted in a pre-formed hole of 2m maximum depth (by micro percussion boring) to extract the soil air by suction. Once the probe has been removed from the hole the sample is transferred to a glass phial and sealed.

R 2 Recommendations on the Principles of Design

R 2-1 Geotechnical Principles in the Design of Landfills

1 General

The various design stages, such as the preparation of plans for approval or implementation, require a geotechnical investigation in accordance with R 1-1 with presentation and assessment of the morphological, hydrological, geological and hydrogeological conditions.

In preparing the design, the following points should be given particular consideration:

- waste management standards relating to the type and quantity of wastes;
- requirements concerning the overall concept of safety for the controlled landfill as an engineered structure;
- extent and duration of site control measures for leachate and landfill gas;
- phased development of the site;
- area available;
- site layout requirements;
- subsequent land use and restoration programme;
- other constraints.

The design, which should be prepared by a qualified geotechnical expert, must take account of factors which are important to the construction, operation and closure of the landfill.

2 Design

All design criteria and requirements relating to the proposed structure, and the safety concept on which these are based, must be set out in the design report.

A risk assessment should be carried out to define the degree of security attainable in relation to the various elements of the design and their interaction. Attention should be given to potential contaminant flow paths.

Design options should be set out and evaluated. All standards and other regulations on which the design is based must be listed, together with any corresponding preliminary decrees, legislation and literature used.

In preparing the design and contract documentation, allowance should be made for the fact that individual structural components may be constructed and checked in sections for convenience and in accordance with the principles of quality assurance (R 5-1, section 3).

In preparing the geotechnical design the following aspects or safety components (barriers) should be considered:

- the subsoil as a site barrier in accordance with R 1-1;
- the basal lining system in accordance with R 2-3;

R 2-1 Geotechnical Design of Landfills

- capping system in accordance with R 2-4;
- the waste body (including internal landfill gas and leachate management systems) in accordance with R 2-6;
- the capping system;
- the disposal site environment;
- geotechnical aspects of site operation;
- geotechnical concerns regarding the removal of gas and leachate from the landfill;
- geotechnical aspects of restoration/recultivation;
- site closure and subsequent aftercare measures;
- supervision and long-term monitoring.

The safety elements must be addressed in terms of their independent function, their mutual influence and overall effect on safety. The implications of failure of one or other of the safety elements must also be addressed in respect of long-term behaviour of the landfill. Necessary remedial measures must be described. The relevant method of site operation must be indicated.

The design documents should contain the following additional information:

- design and geometry of the landfill components;
- site plan and construction programme;
- programme for construction checking and acceptance of individual sections;
- quality assurance plan in accordance with R 5-1, section 2;
- construction material quantities.

3 Other Concerns

The following points should also be considered in the design with regard to construction, operation and closure of the landfill:

3.1 Construction

- protection of components already constructed; in particular, sealing layers and drainage blankets;
- minimum dimensions required for construction work;
- simple and non-sensitive design and construction;
- climatic conditions;
- availability of construction materials.

3.2 Landfill Operation

- waste placement techniques, phasing and supervision;
- stability analyses of the waste body in accordance with R 2-6;
- settlement and other types of deformation in accordance with R 2-4 and R 2-6, checked by monitoring programmes;
- operating facility buildings and roads;
- gas management and monitoring programmes;

- surface water;
- leachate management and groundwater control.

Any changes in the method of site operation should be checked by a qualified geotechnical expert.

3.3 Closure of the Landfill
- Programme for measures to seal the surface and restore the site to the required afteruse, taking account of settlement, drainage and gas emission;
- supervisory measures (type and frequency of monitoring and reporting, review of the monitoring and control programmes, etc.)

R 2-2 Geotechnical Principles in the Design of Remedial Action

- - in preparation - -

R 2-3 Composite Basal Lining System

1 General
A composite basal lining system may comprise a sequence of drainage blankets, transitional and sealing layers. The sealing layers may themselves be composite when a geomembrane is placed directly over a mineral sealing layer, thus minimizing the impact of local defects and combining the advantages of both materials. An example of such a system incorporating a composite liner is illustrated in Fig. 2-3.1.

2 Structure
The structural design of each element in the composite lining system and their relative levels should be specified. A typical example is given in Fig. 2-3.1. The reference number shown is used to identify the corresponding plans in the structural design and to identify the processes, sampling regime, dimensions and documentation during the course of construction.

3 Functions of the Composite Lining System Elements
The suitability of the mineral sealing material must first be confirmed in accordance with R 3-1 and that of the geomembrane in accordance with [J]. In principle, the functions of each element as a geotechnical barrier are as follows:

Mineral sealing layer
- minimisation of seepage and diffusion, related to the choice of material, compaction and thickness of the layer;

R 2-3 Composite Basal Lining System

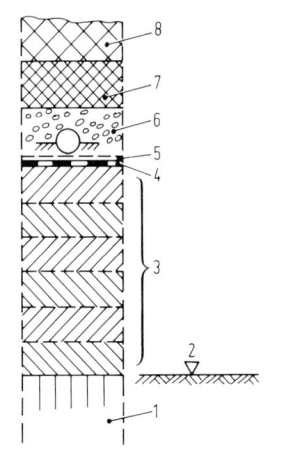

No.	Layer zone
1	subsoil
2	subgrade (in the case of embankment or soil replacement)

3	mineral sealing layer
	    1st lift
	    2nd lift
	    3rd lift
	    last layer in each case

4	geomembrane
5	protective layer
6	drainage blanket
7	transitional layer (if necessary)
8	waste

Fig. 2-3.1
Layer zones and relative levels for a composite basal lining system

- resistance to erosion and water penetration;
- resistance to leachate, related to the swelling clay mineral content;
- heavy metal absorption capacity related to the clay mineral or organic content;
- non-susceptibility to settlement and self-healing ability related to the material's plasticity characteristics, which are determined by clay content and particle size distribution;
- effects of swelling and shrinkage related to hydrogeological conditions.

Geomembrane
- non-susceptibility to settlement related to stress-strain behaviour;
- prevention of leakage;
- long-term chemical resistance depending on the material used and the thickness of the membrane in combination with the mineral sealing layer.

Interface between the sealing layers
- at perforation points in the geomembrane, prevention of the lateral spread of leachate at the interface;
- prevention of any significant water pressure behind the geomembrane;
- sealing effect at the interface related to:
  a) the smooth fine-grained character of the surface of the mineral sealing layer;
  b) the load-dependent deformation behaviour of the geomembrane and the mineral substratum;
  c) effect of change in gradient of the sub-grade related to the load-dependent deformation of the geomembrane.

4 Function of the Layers above the Geomembrane

These layers, arranged one above the other, perform the following roles in the composite system:

Protective layer
- Permanent distribution of concentrated stresses on the geomembrane due to the angularity of the drainage blanket, the protective effect of the geotextile, if any, chemical resistance to leachate and resistance to slippage if appropriate.

Drainage system
- The drainage system permits the collection and removal of leachate from the waste, thereby preventing the build-up of leachate above the sealing system.

Drainage systems can either be discontinuous or preferably, in most cases, a continuous drainage blanket. In both cases it is necessary to take account of the fact that there is a risk of encrustation due to chemical-physical and microbial processes. It is therefore not the filter stability but the risk of encrustation that is of primary importance, and therefore highly transmissive material should be chosen. For mineral materials the following factors should be addressed:

- chemical resistance against leachate;
- leachate permeability of $\geq 1 \times 10^{-3}$ m/s
- stability of the drainage pipes, including pipe bedding and surround, for temperature-dependent conditions;
- ease of maintenance and inspection pipe diameters of at least 300 mm are recommended.

For geosynthetic materials the following points should be considered:

- chemical long-term resistance of the material against leachate;
- appropriate transmissivity under expected in-situ working conditions (mechanical and thermal stress).

The protective layer and the drainage system can, if necessary, be of such a size that unavoidable temperature-induced deformations of the geomembrane are prevented and any effects of frost on the mineral sealing layer avoided.

Transitional layer (if necessary)
- Depending on the design, the transitional layer can have the function of preventing fine-grained waste types from blocking up the drainage blanket. In the construction phase or at the start of the operating phase, it can ensure the necessary surcharge to provide the sealing effect at the interface and prevent frost penetrating the mineral sealing layer. Furthermore, during the exothermic phase of the waste body decomposition, the temperature gradient in the mineral sealing layer can be reduced by provision of a suitable transitional layer.

R 2-3   Composite Basal Lining System   17

5   Design Implications

Material specification may in certain circumstances conflict with the construction methods chosen. For instance, the need for high plasticity in the mineral sealing layer and need to prevent surface rutting. These requirements must be considered in the design in accordance with R 2-1. In addition, the requirements of the overall safety plan must be taken into account in accordance with R 2-1, section 1.

The scope of validation testing necessary for a composite lining system depends on the specific design and stage of construction. This should be laid down in the design and the corresponding suitability check undertaken by a qualified geotechnical expert in accordance with R 2-1, R 3-1 and R 3-3. This should include a recommendation for in-house testing during the construction phase. The regulatory authority will normally define the minimum scope of in-house and external testing required. Recommendations regarding quality assurance in R 2-1 and R 5-2 should also be followed.

R 2-4   Capping System

1   General

The capping system normally comprises a mineral sealing layer, if appropriate incorporating a gas venting layer and drainage system, with a protective covering of subsoil and topsoil (related to the required afteruse). Geomembranes may also be incorporated into the capping system.

2   Structure

The structural design of each element in the capping system and their relative levels should be specified. A typical example is given in Fig. 2-4.1.

3   Functions of the Capping System Elements

The following are the requirements for individual and combined layers within the capping system. There should be adequate slip resistance between the individual bedding planes [K] or friction coefficients should be adjusted to avoid unacceptable tensile stresses.

Waste body and regulating layer
- stability test in accordance with R 2-6;
- minimisation of differential settlement by appropriate waste placement techniques and, if necessary, by surcharging dynamic compaction prior to construction of the capping system;
- preventing the migration of mineral seal material into the waste surface;
- improvement of the pressure distribution of compaction plant;
- if appropriate, permeability to gases of the regulating layer for gas-venting.

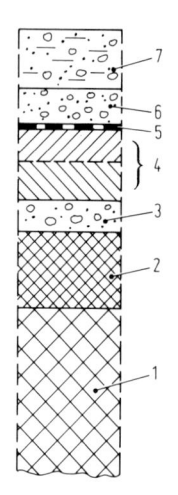

No.	Layer zone
1	waste body
2	regulating layer
3	gas-venting system
4	mineral sealing layer 　1st layer 　2nd layer 　last layer in each case
5	geomembrane
6	drainage system
7	restoration profile 　subsoil 　topsoil

Fig. 2-4.1
Layer zones and subgrade levels for a capping system

Gas-venting layer
- gas emissions picked up over the whole area of the waste body related to layer thickness and permeability to gases;
- resistance to aggressive components in the landfill gas, by use of carbonate-poor material where acid components are present;
- filter stability;
- improvement to the pressure distribution of the regulating layer over the waste body;
- safety from encrustation caused by the deposition of material from landfill gas.

Mineral sealing layer
- suitability check on the sealing material in accordance with R 3-1;
- minimising the formation of leachate, related to the choice of material, compaction, layer thickness and differential settlement;
- non-susceptibility to settlement and self-healing capacity, related to the plasticity of the mineral sealing material determined by clay content and particle size distribution;
- filter stability (i.e. piping and erosion).

Geomembrane
- suitability check on the geomembrane in accordance with [3];
- non-susceptibility to settlement related to stress-strain behaviour;
- prevention of leakage;
- long-term chemical resistance dependent upon the materials used;
- protection against damage to the geomembrane or mineral sealing layer by rodents or root penetration;
- slip resistance in relation to the adjoining layers.

R 2-4 Capping System

Drainage System
- minimisation of hydrostatic pressure on the mineral seal beneath the drainage layer, related to permeability, hydraulic gradient and infiltration rate;
- filter stability;
- slip resistance in embankments [K].

Geotextiles
- improvement of filter stability between adjoining layers;
- resistance to aggressive leachate or landfill gas;
- non-susceptibility to deformation related to differential settlement;
- if appropriate, creation of an area drainage effect related to the type and thickness of the geotextile.

Restoration Profile
- frost protection for the mineral sealing layer;
- erosion protection by vegetation;
- water supply for vegetation in drought periods;
- reduction of infiltration by evapo-transpiration;
- protection of the mineral sealing layer against desiccation;
- reducing the risk of root penetration through the mineral sealing layers depending on the vegetation selected and the thickness and water storage characteristics of the restoration profile.

4 Design Implications

Material specification may in certain circumstances conflict with the construction methods chosen. For instance, the need for high plasticity in the mineral sealing layer related to the need for slip resistance and a low level of shrinkage. These requirements must be taken into account and reconciled in the structural design in accordance with R 2-1 or R 2-2. In addition, the overall safety plan must be considered in accordance with R 2-1.

5 Validation

The scope of validation testing necessary for a capping system depends on the specific design and stage of construction. This should be laid down in the design and the corresponding suitability check undertaken by a qualified geotechnical expert in accordance with R 2-1 and R 3-1. This should include a recommendation for an in-house check during the construction phase. The regulatory authority will normally define the minimum scope of in-house and external testing required. Recommendations regarding quality assurance in R 5-1 and R 5-2 should also be followed.

R 2-5 Health and Safety Considerations at the Contract Stage for Rehabilitation of Contaminated Sites

1 General

At the pre-contract stage for the rehabilitation of contaminated sites the design recommendations in R 2-2 apply mutatis mutandis. The following recommendations should also be taken into account.

A comprehensive list of the aggressive substances found or expected should be prepared, together with corresponding safety measures. This list should cover all media (soil, water, air) relating to the reclamation work. In addition to the preparation of a report on the actual situation at the specific site, it is essential to keep a check on the health of the employees depending on the type of contamination present [4].

2 Invitation to Tender

Specific contractual conditions should point out that additional protective measures must be followed. Rules must be laid down concerning the responsibility for supervision of these protective measures. The tender documents should include a performance specification related to specific objectives clearly indicating the scope of protective measures and associated remuneration. The overall safety plan must be summarised in an annex to the performance specification which must contain, in particular, safety requirements which go beyond normal accident prevention provisions. The following points must be taken into account in drawing up a performance specification for protective measures:

- list of aggressive substances present, given separately for water, soil and air;
- list of corresponding necessary protective measures;
- measures relating to industrial safety, which go beyond what is normal for the building industry, should be dealt with as separate items. Such measures include:
  - requirements which reduce output (for instance daily or frequent decontamination of vehicles);
  - type and scope of personal protection to be used;
  - reservation and operation of sanitary facilities required;
  - working and operating instructions;
  - necessary medical check-ups;
  - measuring programmes to be carried out.

3 Preparation for Work

The scale of medical check-ups required should be laid down in collaboration with experienced specialists (in toxicology or industrial medicine).

The following should also be established:
- emergency planning
- operating instructions

R 2-5 Safety Considerations for Rehabilitation of Contaminated Sites   21

- measurements/measurement techniques,
- reporting procedures.

The following must be taken into account in organizing the construction site:
- definition of the site boundary;
- washrooms, accommodation, rest rooms;
- medical facilities, if appropriate;
- storage and cleaning of equipment;
- fire protection measures;
- a weather station, in certain circumstances;
- clean and dirty sections for personnel and equipment, if necessary.

4 Protective Measures

In principle, all measures which serve to limit emissions are conducive to industrial safety.

4.1 Operational Protective Measures

These include:
- small excavation areas;
- temporary covers.

A check should be made on whether the following are needed:
- vehicle sluices;
- lorry washing facilities;
- transport documentation procedures.

4.2 Personnel Protection

Where there is a conflict between maximum safety and adequate freedom of movement, the following measures are available for protection against skin contact:
- rubber safety boots;
- safety goggles or protective facial shield;
- disposable chemical safety suits;
- plastic coated protective gloves.

The following may be necessary in certain circumstances for respiratory protection:
- dust masks, respiratory device;
- respirators.

In addition, examinations and/or training must be given to ensure that employees are suited from the point of view of:
- health; and
- technical capability.

R 2-6 Waste Body Stability

1 General

In a typical landfill, the waste is built up in layers to form a waste body. The type of waste, its state of placement, and increases in stress caused by filling will influence the extent vertical and horizontal deformation of the waste body. Such deformation may be intensified by long-term bio-chemical changes within the waste. In addition, settlement and lateral deformation can, in fine-grained and impermeable areas of the waste body, be accompanied by a reduction in shear strength due to pore water and pore gas pressure.

When considering landfill stability and deformation problems it is convenient to identify two basic aspects (see Fig. 2-6.1):

- "External" stability refers to the strength and deformation behaviour of the waste influencing potential failure zones for all slopes (temporary and final) both during and after the operational phase. If waste with too low strength is placed within these zones, initial lateral deformation in the lower slope region and settlement in the upper region can progressively lead to slope failure.
- "Internal" stability relates to the placement of waste within the zone which does not influence external stability as defined above. It is within this zone that wastes with low strength can be placed, subject to consideration being given to operational stability. It is also necessary to maintain a safe distance from the surface of the waste body (which should include the slope safety zone) so that any stresses on the capping system caused by differential settlement are kept within permitted limits.

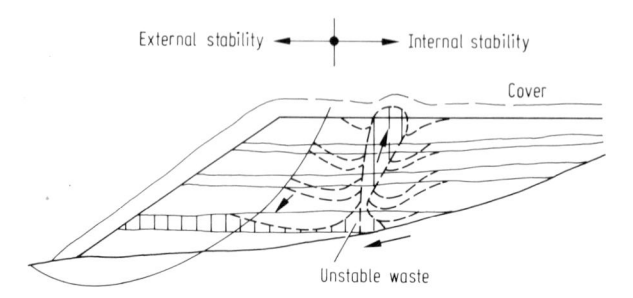

Fig. 2-6.1
External and internal stability [5]

2 Settlement and Lateral Deformation

The compatibility of relative strain between the waste and individual elements of the lining system and/or other structures within the waste body must be verified.

The following definitions relate solely to the mechanical properties of waste.

2.1 Soil-like Waste

In the case of soil-like granular waste (defined as particulate materials for which Soil Mechanics principles are applicable which may include sludges, ashes, excavation material, etc.) the settlement and lateral deformation behaviour should be deduced from stress deformation tests undertaken in accordance with R 3-6; the actual deformation behaviour of the waste body should be defined by field measurements.

2.2 Other Waste

In the case of non soil-like wastes, (defined as materials for which Soil Mechanics principles are not applicable, and which may include municipal waste and mixed waste) a realistic prediction of settlement and lateral deformation is currently possible solely on the basis of monitoring waste bodies of similar construction. Due to the relatively long periods for which disposal sites are established there is generally sufficient time available to measure the time-induced deformation behaviour of the waste body [L] and to therefore draw conclusions on the implications for stresses on the capping system.

It is recommended that on reaching one third and two thirds of the proposed ultimate height of the site, at least one depth settlement gauge should be installed for every 3 ha of site area and measurements should be taken at quarterly intervals. From the settlement behaviour, measured over a period of time, it is possible to determine the deformation behaviour of the surface of the waste body to provide realistic forecasts. In designing the final landform, and calculating the available capacity, account should be taken of settlement of the waste body. Detailed records of the type and location of different waste types are essential, if correct understanding of the settlement behaviour of the waste is to be achieved.

The following method is suggested for estimating the settlement of the waste body:

- ascertain the overall settlement in relation to depth of waste placement, placement technique and waste composition with approximate values for deformability parameters;
- monitor settlement with time by calculating gas production or other related biochemical characteristics.

To minimize significant differential settlement which may result in unacceptable strain on the capping system the following situations should be avoided:

- wastes with widely differing composition or different consistency and settlement capacity being placed close to the waste surface;
- the upper sections of the waste body forming the subgrade to the cap varying widely in age or pre-loading.

Lateral deformations in the central slope region are an indication of the characteristic deformation behaviour caused by progressive loading due to increasing the height of waste. On the other hand, significant lateral displacement in the lower slope region, in conjunction with significant settlement in the upper slope region can indicate that the strength of the waste body in the slope zone is mobilized to a significant degree. In this case inclinometer measurements may be appropriate in addition to surface geodetic surveying.

3 Stability Analysis

3.1 Slope Failure

In the case of soil-like waste the stability of a slope is generally assessed using conventional analysis [K]. For this purpose it is necessary to establish the unit weight of the waste body and the relevant strength parameters according to the age of the waste. Pore pressure variations which could have a significant influence on slope stability must be carefully taken into account in the calculations. In the case of other waste, these parameters may be determined either by using appropriate large shear box apparatus or triaxial compression cells or by performing large scale field tests. Particular attention should be given to the effect of age and decomposition of the waste on its mechanical properties.

All observations on the behaviour of the shape of the waste body (in terms of settlement and lateral deformation) should be passed to the appointed qualified geotechnical expert who is then responsible for the continuous assessment of site stability, both during and after the operational phases of the landfill.

3.2 Additional Spreading Induced Stresses

In addition to conventional rigid body analysis, attention should be focused on stress concentration due to interaction between waste and lining system. Vertical settlement can induce associated tensile stresses in the lining elements [M].

3.3 Overall Site Stability (Ground Failure)

The overall stability of the site (waste and surrounding ground) must be assessed by conventional stability and bearing capacity analyses [K].

4 Structural Design Notes

The wastes to be placed in the waste body and the placement methods required to adhere to calculation parameters should be laid down in an operating plan, in agreement with the site operator, when the structural design is prepared in accordance with R 2-1.

The geotechnical design provides the basis for defining the following requirements to be imposed in order to maintain stability:
- the permitted wastes in accordance with the suitability check under R 3-6;
- the placement of the wastes and appropriate checks under R 5-4 as the waste body is built up in the region affecting stability (both within the slope zone and the upper part of the tip close to the surface zone).

R 3 Recommendations on Suitability Testing
R 3-1 Suitability Testing of Mineral Capping and Basal Seals
1 General
The suitability of all materials used in the construction and rehabilitation of landfills and contaminated sites must be proved. The suitability of minerals to be used to seal landfills generally should be proved separately for each specific application. The number of random samples required for this should be defined by the qualified geotechnical expert.

The following investigations should be carried out:
- determination of the nature and composition of the sealing material;
- establishment of the placement criteria;
- determination of the characteristics of the sealing material when in place.

2 Tests Relating to Soil-Physical Classification
The materials proposed for use as a mineral seal should be classified in accordance with [N]. This also applies to in-situ soils whose sealing properties are to be used in the design.

To describe mineral sealing materials, the characteristic parameters set out in Table 3-1.1 should be determined by laboratory tests on a sufficient number of representative samples. The range of scatter normally to be expected for natural deposits should also be documented and made available as a basis for preparation of the design in accordance with R 2-1.

The tests listed in Table 3-1.1 in respect of grain-size distribution, plasticity, shrinkage limit, water intake and calcium carbonate content provide an initial indication of the type of clay mineral and cementing agent present (e.g. carbonate). If necessary, more detailed information on the mineral content of the fine-grained fraction can be obtained from a semi-quantitative mineralogical analysis in which the clay minerals montmorillonite, mixed layer, illite or kaolinite, and their respective dominance, are determined. In special cases a quantitative mineralogical analysis may be necessary.

Table 3-1.1
Characteristic parameters for the soil-physical classification of mineral sealing materials for landfills

Parameter	to be determined in accordance with [0]
Grain size distribution	[01]
Consistency limits (liquid limit, plastic limit, shrinkage limit, plasticity index, consistency index)	[02]
Organic constituent content (ignition loss or wet oxidation)	[03]
Grain density	[04]
Calcium carbonate content	[05]
Water intake	[06]
Moisture content	[07]
Density	[08]

3 Determination of Placement Criteria

Key placement criteria for mineral seals are density and moisture content. The density achievable as a function of the moisture content should be determined by the Proctor test in accordance with [P]. The degree of compaction required for placement and the placement moisture content should be determined in association with permeability tests.

A field scale trial placement and compaction, in accordance with R 3-5, should be used to validate the suitability of a mineral sealing material selected on the basis of the laboratory results. In addition, the lift thickness proposed for layered placement, the required number of passes to be made with compaction plant, the suitability of the earthworks machinery and the effectiveness of the proposed control checks should be examined. In individual cases, the qualified geotechnical expert should design, supervise and review the execution of trial compactions. This should also include sampling [B] and the execution of laboratory tests (Table 3-1.1), in particular Proctor compaction and permeability tests.

4 Stress Deformation Behaviour

In order to assess the deformation behaviour and swelling characteristics of mineral sealing materials for stability calculation purposes, the compressibility, swelling behaviour and shear strength should be determined (Table 3-1.2). For triaxial compression tests, direct shear tests and unixial compression tests the stress deformation pattern should be recorded.

To achieve this, samples should either be taken from trial areas [B] or test specimens prepared in the laboratory at the relevant placement values (Proctor density and moisture content).

R 3-1 Mineral Seals

Table 3-1.2
Tests to assess the stress deformation behaviour of sealing materials [Q]

Type of test	to be conducted in accordance with
Compression test	[Q1]
Swelling	[Q2]
Triaxial compression test or direct shear test	[Q3]
Uniaxial compression test	[Q4]

5 Permeability Testing

The permeability behaviour of mineral sealing materials should be examined in the laboratory [R]. The following supplementary or additional rules apply to the sealing of landfills.

The effect of chemical or physical reactions between a test liquid and the sealing material on permeability behaviour should be assessed using an appropriate test facility.

When testing the permeability of soils, test specimen dimensions should be selected by reference to the grain size distribution of the material [S].

The triaxial compression test with constant hydraulic gradient is recommended as the preferred method for examining the permeability of fine-grained soils. An isotropic stress condition is imposed on the sample; as a rule, the pressure should be set at 0.3 bars above the pore-water pressure. For testing mineral sealing materials with a maximum particle size of ≥ 20 mm the use of other methods in accordance with [R] is recommended.

Demineralised water should be made to flow through the sample from bottom to top, the aim being to achieve maximum saturation. The degree of saturation achieved should be determined at the end of the test. The temperature of the water should be kept constant and correction for temperature made in accordance with [R].

The permeability coefficient k should be determined with a hydraulic gradient $i = 30$. In certain cases it may be necessary to conduct additional tests at the hydraulic gradient anticipated in the field.

In any event, the permeability test should be continued until the trend in permeability coefficient remains more or less constant. The permeability coefficients should be set out in graphical form as a function of the test duration and/or as a function of the number of pore volumes displaced.

An assessment of permeability behaviour based purely on permeability coefficients found using water is not always sufficient. It is therefore recommended that additional permeability tests be carried out preferably using leachate from the landfill (particularly in the case of monodisposal-landfills).

If empirical data for the composition of leachate from other landfills are available, these should be taken into account in selecting appropriate test liquids. If no empirical data are available, the following test liquids could be used:
- distilled water for comparison;
- strong acids, e.g. hydrochloric acid (pH ≤ 3);
- strong alkalis, e.g. a caustic soda solution (pH ≥ 11);
- metallic salt solutions (conductivity $\geq 10{,}000\ \mu S/cm$);
- hexane (to the limit of solubility);
- halogenated volatile hydrocarbons (to the limit of solubility);
- acetone ($\leq 10\%$);

Other liquids, such as phenols (100 g/l) or benzole (to the limit of solubility) can also be used, but they require special precautions in the laboratory.

The permeability test must be continued until sample and test liquid form a "stable system", i.e. the liquid flowing through is not found to cause any changes in the sealing material or the liquid. One way of assessing this is by chemical analysis of the liquid before and after it flows through the material (for example, constant pH and conductivity data).

6 Assessment of Change in Long-Term Chemical Stability

The effects of leachate on the long-term behaviour of mineral sealing materials, especially the clay minerals contained in the finest particles, can be assessed by undertaking tests to identify any change in characteristic soil-physical values of the material and/or the finest particle fraction.

If, for instance, in the water intake test [T], another type of test liquid is used in place of tap water, it is possible to make a qualitative statement about the effects of the liquid on the chemical stability of the mineral sealing material (on the basis of any change in liquid intake and the trend over a period of time). The same applies to the use of leachate.

If no empirical data are available on the composition of the leachate, the test liquids listed in section 5 should be used.

Chemical effects on mineral sealing materials may also be assessed using the following tests:
- change in grain-size distribution after removal of any mobile cementing agent (R 3-3) and/or after treating with a leachate or test liquid;
- chemical characterisation of the cementing agent (e.g. calcite, dolomite or iron oxides);
- change in swelling characteristics of the material after treatment with a leachate or test liquid;

- change in plasticity after removal of any mobile cementing agent and/or after treatment with a leachate or test liquid;
- change in water intake of the sealing material after removal of any mobile cementing agent (R 3-3) and/or after treatment with a leachate or test liquid.

If additional information is required concerning the influence of leachate, corresponding clay mineralogical tests should be conducted in accordance with R 3-3.

R 3-2 Suitability Testing for Mineral Sealing Compounds

1   General

This technical recommendation lays down a methodology for testing the suitability of mineral sealing compounds used for the vertical sealing of landfills and contaminated sites. It applies to all methods of construction in which the in situ soil is displaced (e.g. thin wall) or dug out (e.g. diaphragm wall) and replaced by a mineral sealing compound. The principal construction methods are as follows:

Thin wall:	A steel profile is vibrated in. When the profile is pulled out the resulting space is filled with a sealing compound under pressure.
Diaphragm wall: one-phase method	Soil is dug out under the protection of a hardening bentonite-cement or other suspension which remains in the cavity and slowly sets.
Diaphragm wall: two-phase method	Soil is dug out under the protection of bentonite or other suspension, the sealing compound is introduced by the tremie method and the suspension is simultaneously displaced.

In certain circumstances a geomembrane may be introduced in a diaphragm wall (see Appendix).

This guidance does not apply to methods in which the pore space of the in situ soil is reduced or filled by grouting techniques.

The suitability of mineral sealing compounds must be demonstrated in each individual case by suitability tests. These must be undertaken only by laboratories with the appropriate equipment and expertise. All test methods indicated hereafter are for guidance.

As a general rule, the following should be determined:

- composition and characteristics of the individual components of the mix;
- characteristics of the fresh sealing compound;
- workability and solidification behaviour of the sealing compound;
- strength and stress deformation behaviour of the hardened sealing compound;
- permeability of the hardened sealing compound (R 3-1);
- unit mass and water content;
- physico-chemical properties related to specific requirements.

The water intake test, [T], can be used to identify the material characteristics of the sealing compound [6]. The sealing compound characteristics established in suitability tests should be checked on site by quality assurance techniques in accordance with R 5-3.

2 Composition and Manufacture of the Sealing Compounds and Preparation of Samples

Sealing compounds generally comprise the following components:
- colloids (such as bentonite or others);
- hydraulic binding agent;
- mineral fillers;
- water and additives if necessary.

The characteristics of the individual components must be demonstrated by test certificates issued by the manufacturer or by corresponding tests undertaken by the laboratory conducting the suitability programme. The following information is relevant, particularly when bentonite is used:

Bentonite:
- description in accordance with [U];
- water intake [T];
- montmorillonite efficiency [7].

Hydraulic Binding Agent:
- for cement particulars in accordance with [V].

Mineral Fillers:
- soil-physical description (grain shape, grain-size distribution, organic constituent content, grain density, calcium carbonate content, water intake, moisture content);
- mineralogical and chemical description;
- non toxicity.

Water:
- particulars in accordance with [W].

Additives:
- type of substance and its action;
- quantity of dry substance;
- test certificates.

The effectiveness of individual components used in the suitability test must not be reduced by an extended storage period. The date of manufacture should therefore be indicated.

The composition of a sealing compound should be defined on the basis of the suitability tests. Since local materials are often used in mineral sealing compounds, these must also be considered in the suitability tests. Any incorporation of excavation debris into the compound during wall construction should also be considered in the test programme if appropriate.

R 3-2 Mineral Sealing Compounds

As a rule, the composition of a sealing compound should be specified per cubic metre of fresh compound.

Mix proportions should be specified in kg per cubic metre of fresh sealing compound in the sequence colloid/hydraulic binding agent/mineral fillers/water, e.g. 40/200/100/880 mix. In addition, the unit masses of the component materials and of the mixture should be specified. To make the sealing compound in the laboratory, the bentonite should first be prepared in accordance with [U] and checked after an expansion time of 24 hours. Depending on site specific conditions, this time period may need to be modified subject to agreement between the contractor and the laboratory.

The proportions of hydraulic binding agent, mineral fillers and any chemical additives are added to the expanded bentonite suspension by further mixing processes in the sequence proposed for the specific project. A mixing time of 5 minutes at 1,200 rpm should generally be adhered to. An example of the mixing equipment used, in this case a multiblade turbo propeller, is given in Fig. 3-2.1. If the mixing container illustrated in Fig. 3-2.1 is used, it should be filled with at least 15 litres of sealing compound.

All samples should be prepared in one batch.

The characteristics of the fresh sealing compound should be tested in accordance with section 3 below. Testing should take place immediately after mixing. If no empirical data are available, the test should be repeated after a one hour resting period.

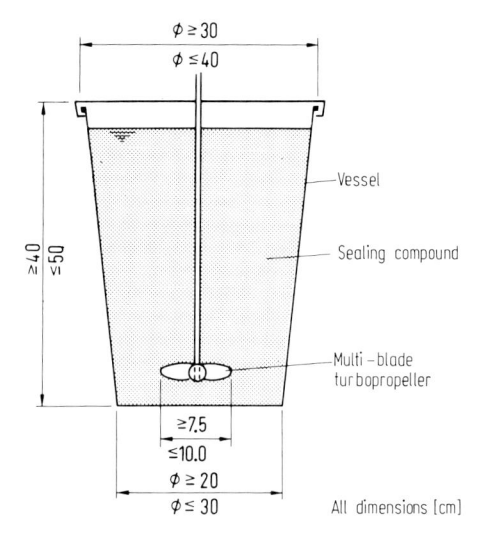

Fig. 3-2.1
Manufacture of the sealing compound in the laboratory, mixing container and mixing blades

When testing to define the mechanical characteristics and permeability of the hardened sealing compound (in accordance with sections 5 and 6 below), appropriate sample sizes should be chosen. For permeability testing the fluid sealing compound should be poured into short lengths of pipe 10 cm in diameter and at least 12 cm long. These dimensions can be reduced by reference to the grain-size distribution of the product, particularly for mechanical testing, where an h/d ratio of up to 2 can be considered.

All samples should then be stored under water. Throughout the storage period, the water must be maintained at a temperature of 18 °C +/− 2 °C.

3 Testing the Characteristics of the Fresh Sealing Compound

The characteristics of the colloid suspension in the two-phase method, or of the sealing compound in the one-phase method, determine the stability of the open cavity during construction and must therefore be determined. In addition, the rheological characteristics give an indication of the workability of the sealing compound.

In particular, the characteristic values listed in Table 3-2.1 should be determined. Further explanatory notes on the test equipment may be found in [U].

4 Tests on Workability and Hardening Behaviour

In order to assess the workability and hardening behaviour of mineral sealing compounds, agitation tests may be conducted. In the agitation test the sealing compound is agitated for five minutes and then left to rest for 15 minutes. This process continues

Table 3-2.1
Measurement of the characteristics of the bentonite suspension or the fresh sealing compound

Characteristic value	Measuring equipment
Unit mass (kg/m^3)	Mud balance Pycnometer
Run-out time(s)	Marsh funnel
Filtrate water loss (m^3)	Filter press (API) 700 kPa 450 s for bentonite slurries (*)
Yield value (Pa)	Sphere harp Pendulum apparatus Viscosimeter
pH value (−)	pH meter pH paper
Bleeding (%)	Graduate vessel

(*) For other slurries, or when bentonite cement mixes are considered (one phase diaphram wall), other appropriate equipment and operating parameters should be used.

for the estimated length of time required to place the sealing compound on site. The agitation tests should be conducted for a minimum of eight hours. The dimensions of the mixing device and vessel may deviate from those given in section 2 above. It is important to ensure that all the sealing compound is kept moving while the test is being conducted. The mixing device must revolve at around 500 +/− 200 rpm. Every two hours the run-out time, yield value and filtrate water loss should be determined and the sealing compound poured into sections of pipe, and then stored in water at 18 °C +/− 2 °C. After a storage period of 28 or 56 days, the samples are tested for uniaxial compressive strength and the results set out in graphical form as a function of the agitation time.

Additional tests on hardening and setting behaviour using a laboratory vane may be appropriate.

5 Strength and Stress Deformation Behaviour

In order to assess the stress deformation behaviour, the uniaxial compressive strength of three hardened samples should be determined after 14, 28 and 56 days in accordance with [Q]. The h/d ratio for the sample should be in the range 1.0 to 2.0. The machine speed should be set at a value corresponding to around 1 % vertical strain per minute. The complete stress strain diagram should be included in the test result. The uniaxial compressive strength should be set out in graphical form as a function of sample age.

For more detailed information on deformation behaviour, or for strength or creep assessment, conventional soil mechanics compressibility tests or triaxial tests can be conducted. This applies, for example, if as part of the design, load bearing sheet pile walls, pre-cast reinforced concrete sections or similar are to be incorporated into the upper part of the wall or if the sealing wall is designed to resist deformation stresses (e.g. subsidence, one-sided load imposed adjacent to the wall, etc.).

6 Permeability Testing

The permeability behaviour of mineral sealing compounds should be tested after being stored in water for 28 or 56 days. For water permeability testing, R 3-1, section 5 applies, taking account of the sample dimensions recommended in section 2 above.

The permeability behaviour of mineral sealing compounds should not be based purely on results obtained using water. Permeability tests should also be conducted using representative leachate samples or test liquids (synthetic leachates). In determining the composition of test liquids, the highest proven concentrations of each aggressive substance should be considered. The composition of a suitable test liquid should be defined by the geotechnical and chemical experts.

R 3-3   Clay Mineralogical Characterisation of Mineral Basal Seals

1   General

In order to describe the mineralogical composition of the clay mineral fraction of sealing materials, and to ascertain the potential influences on permeability of chemical interaction with leachate, clay mineralogical and geochemical tests are necessary in addition to the suitability tests described in R 3-1.

The tests may be conducted either on fresh samples or on pre-contaminated sample material. The procedures described are a modified form of the routine methods used in sediment petrography.

2   Grain Size Distribution

In order to define the finest fraction (> 0.0022 mm) the material must first be dissolved chemically (e.g. with NHCl solution) to avoid determining a purely random aggregate condition. By treating the samples several times with 0.1 molar solutions of ethylene diamine tetra-acetic acid (= EDTA; e.g. Titriplex) having a pH value of between 4.5 and 8, the Ca and Ca/Mg carbonates occurring as binding agents are gently dissolved, and the Ca and Mg ions adsorbed on the clay mineral surfaces are exchanged for Na ions in the EDTA. This also allows an optimum degree of dispersion. Using the samples thus prepared, a sedimentation test is conducted in accordance with [O].

3   Clay Mineralogical Testing

3.1   Determination of the Ion Exchange Capacity

By applying monovalent or bivalent metal cations or NH_4+ ions to the clay mineral surfaces of a soil, and subsequent quantitive re-exchanging of the fixed cations, the proportion of swelling clay minerals in the sample can be determined [7], [8].

A test sample weighing a few grammes is sufficient for these analyses (Table 3-3.1). It is necessary to treat the sample several times with an exchange solution, (e.g. 3 N ammonium acetate solution) and to remove excess ammonium acetate before the Kjehldahl analysis [8] by washing the sample several times.

R 3-3 Clay Mineralogical Characterisation

Table 3-3.1
Maximum test size and ion exchange capacity for various clay minerals for the *Kjehldahl* procedure [8]

	max. ion exchange capacity (mval/100g)	max. test size (g)
Kaolinite/fireclay	25	10
Illite	50	5
Halloysite	100	2
Attapulgite	100	2
Mixed layer minerals	100	2
Montmorillonite	200	1

3.2 X-ray Defraction Analysis (XRD)

The X-ray defraction analysis of the various clay mineral constituents should be conducted using only elutriated, finest grain material ($< 2 \,\mu$m) free of any cementing agent. Since only about 10 to 40 mg of material is required for the preparation of a natural and glycerine saturated (and also ethylene glycol saturated, if necessary) sample, a simplified sedimentation procedure (e.g. pipette method) is adequate for separation of the finest fraction. By considering the proportion of the principal clay minerals (such as illite, montmorillonite, kaolinite and chlorite) in the finest fraction, semi-quantitative details of the clay mineral content can also be obtained by means of XRD analysis. The chemically active clay minerals occur almost exclusively in the fraction $< 2 \,\mu$m; kaolinite usually also occurs in the fraction between 2 and 6.3 μm; montmorillonite is concentrated almost entirely in the fraction $< 0.2 \,\mu$m.

3.3 Water Intake

In addition to the requirements of R 3-1, the water intake test can be conducted on the finest particle fraction free of cementing material [T]. By reference to Fig. 3-3.1, errors in estimating the clay mineral proportions can be reduced and information can be derived concerning the overall swelling properties. The test should last 24 hours.

4 Carbonate Ratio

When determining the calcium carbonate content it is necessary to distinguish between the calcitic and dolomitic cementing agents because of their varying chemical stability. The calcite and dolomite content is most simply determined by titrimetry using a Ca titration and a Ca-Mg cumulative titration [T]. It is also possible to detect iron, which may also form a carbonate cementing agent in the hydrochloric acid solu-

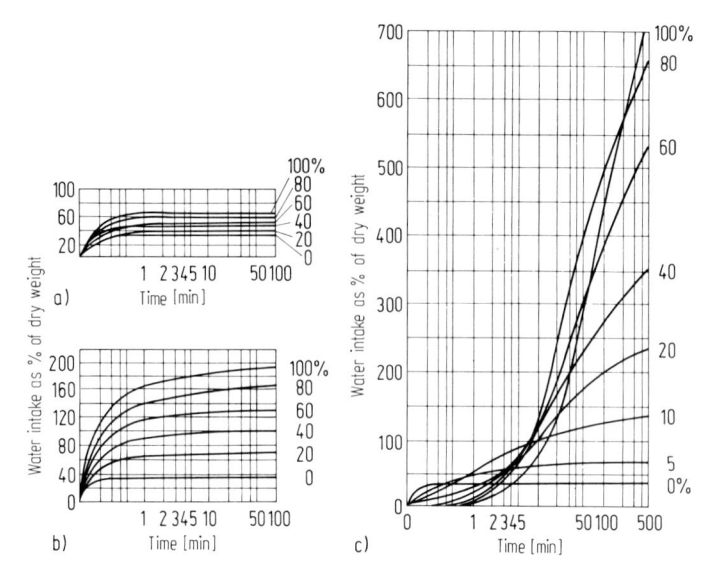

Fig. 3-3.1
Water intake of various clay mineral/ground quartz mixtures (in accordance with *Pichler*) [10]
Proportions given in % weight of the clay ratio
a) Quartz kaolinite
b) Quartz Ca bentonite
c) Quartz montmorillonite

tion used for these titration tests. As an alternative to titration, use may be made of atomic absorption analysis (AAS). The calcite/dolomite ratio can also be determined sufficiently accurately by XRD [9].

R 3-4 Chemical Impact of Leachate on Mineral Sealing Materials

1 General

Due to the varying chemical stability of clay minerals and the complex chemical nature of landfill leachate, it is still not possible to draw up any generally valid criteria for the potential chemical effects of leachate on mineral sealing materials.

The leachate parameters hitherto routinely adopted, such as:
- pH value
- electrical conductivity
- content of cations and anions
- acid capacity up to pH 4.3; base capacity up to pH 8.2
- carbonate hardness

R 3-4 Chemical Impact of Leachate

- COD: chemical oxygen demand
- BOD: biological oxygen demand
- evaporation residue
- ignition loss
- TOC: total organic carbon

are only of limited use for characterising the potential chemical impact on mineral sealing materials. The specific points relating to leachate are summarised below.

2 Chemical Characterisation of Leachate

2.1 Observations on Site

- Temperature, pH value and electrical conductivity as a function of seasonal fluctuations or the age of waste within the landfill;
- temperature, turbidity and smell.

2.2 Characterisation of the Organic Components of Leachate

- extractable organic halogen compounds;
- organic halogen compounds;
- highly volatile to semi-volatile aromatic and aliphatic hydrocarbons;
- total hydrocarbons;
- highly volatile halogenated hydrocarbons;
- phenols;
- PAH;
- PCB;
- chlorinated phenols.

Using these parameters, the effect of hydrocarbon compounds on the mineral sealing material can be estimated and the solubility in water of organic leachate constituents can be assessed. The potential interactions between clay components and leachate can only be assessed if these leachate parameters and the mineral characteristics quoted in R 3-3 are known.

R 3-5 Field Trials for Mineral Basal Seals and Caps

1 Design Principles

Preparation of a field trial for the construction of a mineral seal for a landfill should be regarded as a large-scale suitability test in which the external and in-house examiner should be involved. A field trial demonstrates the following points using sealing material tested in the laboratory for suitability in accordance with R 3-1:

- suitability of the material under site conditions;
- suitability of the methods of extraction, treatment and preparation;
- suitability of the methods of placement and compaction;

- adherence to the requirements for permeability, water content and density of the mineral sealing material on a large-scale basis;
- establishment of reference parameters for quality assurance.

The preparation of a field trial forms part of the quality assurance programme for construction of a landfill. It is recommended that the field trial itself should not be incorporated into the final sealing layer.

The field trial should be prepared in good time before the sealing layer is constructed. The subgrade for the basal seal must be carefully prepared so that it meets all the requirements for the subsequent foundation for the landfill. The subgrade of the surface seal must correspondingly meet all the requirements for the proposed capping system.

For mineral seals with a gradient steeper than 1:4, a field trial inclined according to the design should be prepared in addition to one which is approximately horizontal (Fig. 3-5.1).

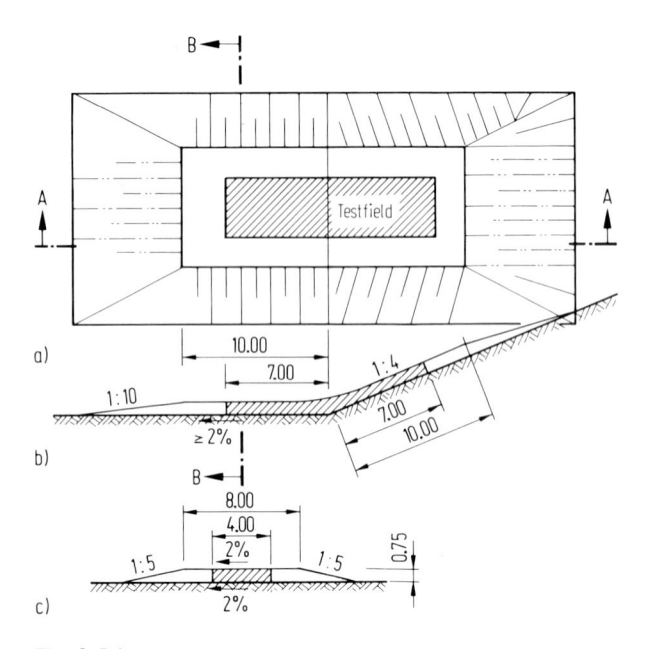

Fig. 3-5.1
Example of a sample field trial with horizontal section and 1:4 gradient section.
a) Plan; b) Section A–A; c) Section B–B

        Prepared subgrade
        Test field
          $\geq$ 3 lifts, compacted
          $\geq$ 75 cm thickness

R 3-5 Field Trials for Mineral Seals

The field trial should be prepared using the material to be used for the mineral basal seal. It is possible for the mineral seal to be made up of several components if a special mixing process is used. For conditioning the mineral construction materials and for their placement and compaction, the same equipment, methods and working conditions (e.g. mineral composition, moisture content, duration of treatment, weather conditions, etc.) should be used for preparing a field trial as those proposed for placement of the mineral seal. If significant variations are to be expected in the characteristics of the relevant soil types or equipment to be used, it may be necessary to prepare several field trials with different initial materials and different equipment, or one larger field trial comprising several discrete test areas. This applies mutatis mutandis for mixtures of minerals, particularly if the final composition is to be defined on the basis of the results of the compaction trials.

The preparation of the field trial should be designed, supervised and then assessed by a qualified geotechnical expert. In individual cases deviations from, or supplements to current recommendations should be checked to ensure they are appropriate for particular methods or sites.

In planning the field trial programme the frequency and scope of sampling and testing should be specified.

All measurements and observations associated with the field trial, together with the conduct of field and laboratory tests and all other survey results, must be carefully and fully described, recorded and documented.

2 Execution of the Work

The dimensions of the field trial should relate to the design of the seal, the earthworks equipment used and the requirements of the proposed field tests, samples, measurements and observations. The gradient of the subgrade on which the field trial is placed must correspond to that of the actual structure.

The dimensions of the field trial should be such that, after removal of the perimeter zones, a sufficiently large, representative area is available having regard to the size of the compaction equipment used. At least 3 lifts 20 to 25 cm thick (in the compacted state) should be placed and compacted, regardless of the final thickness of the sealing layer.

The subgrade on which the field trial is placed should be surveyed geodetically by position and elevation before placement begins. The condition and preparation of the subgrade should be documented. While the field trial is being constructed and the tests are being undertaken, daily reports should be produced summarising weather conditions (temperature, rainfall, insolation or cloud cover and wind conditions). Each completed lift should be surveyed geodetically, as should the completed field trial; position, elevation, gradient and level should be determined and the condition of the field trial described.

If the measuring grid is kept constant, it is possible to provide non-destructive proof of the thickness of the liner by comparing surveyed levels before and after placement (this ignores the effect of subsidence).

For each field trial the following should be recorded:
- origin, type and condition of the mineral sealing material;
- methods used for extraction, transport, treatment and placement of the material;
- type, operating principle, weight and major dimensions of the equipment used;
- diameter and length of the roller; operating weight with and without ballast; operating speed of the compaction equipment; frequency and energy of vibrating rollers; length, cross-sectional area and arrangement of the studs used to achieve a kneading effect in the case of sheep's foot rollers;
- number of passes with the roller, indicated separately for each type where various types of roller are used;
- type, dimensions and characteristic values of any soil conditioning equipment used;
- number of rotavations and operating speed of the machine;
- methods used for breaking up clods of soil, maximum permissible clod size and degree of breaking up achieved;
- method of checking and, if necessary, correcting the moisture content of the soil to be placed, origin of the added water, period elapsing between distribution of the water and the commencement of compaction;
- thickness of the lift before and after compaction;
- if necessary, addition of bentonite, powdered clay or other additives; quantities used, method of batching, number of rotavations or duration of mixing in the pressure mixer.

The individual lifts should be protected against atmospheric effects by providing a ballasted covering sheet. In some cases, trafficability tests on the completed mineral subgrade may be appropriate on completion of the investigation.

As the final check on the homogeneity of the seal, the field trial should be dismantled using an excavator. Any inhomogeneities found which were not detected in the preceding investigation programme should be examined.

3 Sampling, Field and Laboratory Tests

In order to determine density and moisture content, and thereby define the degree of compaction and permeability achieved, sufficient samples should be taken in accordance with [B] from at least 3 locations after completion of each layer and also after completion of the whole field trial. In addition, at least one sample should be taken from each interface between lifts for the purpose of measuring permeability in the laboratory. Grain-size distribution and consistency limits [X], as well as other appropriate classification tests (such as water intake tests [T]) are to be determined on one disturbed sample per lift. In addition, water intake and moisture content should be determined for 5 random samples per lift in order to confirm material homogeneity. Further sampling to examine compressibility, expansion behaviour, shear strength or other mechanical properties may be appropriate in individual cases (see R 5–2). All laboratory tests should be conducted in accordance with R 3-1.

If there is provision for the measurement of density and water content using the nuclear density meter the instrument should be calibrated on the field trial and by reference to laboratory measurements on special samples.

All field trial results should be reviewed in accordance with section 1 above. If the minimum requirements for permeability, density, homogeneity, etc., were not fully achieved, then proposed measures to satisfy these requirements in the final structure should be prepared.

Field trial data should be compared with the initial material suitability data obtained from R 3-1.

R 3-6 Suitability Testing of Waste Substances for Placement in the External Stability Zones of Landfills

1 General

The assessment of the stability of landfills (R 2-6) requires suitability tests on the mechanical properties of non-treated or conditioned wastes. These tests establish the requirements for identification of the waste for the site operator's acceptance and quality checks (R 5-4).

For the suitability test for placement of waste in the internal stability zone of the waste body (see R 2-6), refer to the corresponding methods given in [11].

Suitability testing for placement in the external stability zones (R 2-6) (i.e. in the slope and upper surface zones of the landfill) is discussed below.

2 Geotechnical Description and Classification of Waste

Waste substances approved for placement within landfills, in accordance with waste regulations, should be described and classified in relation to mechanical behaviour of the waste body. For this purpose waste types are divided into two groups:
- soil-like waste, defined as granular waste, for which conventional Soil Mechanics theory is applicable;
- other waste (non-sorted municipal waste, etc.)

2.1 Soil-like Waste

Soil-like waste which can be compared with soil groups [N] in respect of composition and geotechnical behaviour, (e.g. fine-grained clays, mixed grained and coarse grained materials with mineral and/or other constituents) are assessed in accordance with Table 3-6.1.

Table 3-6.1
Geotechnical classification of soil-like wastes

Parameter	to be determined in accordance with [Y]
Moisture content	[Y1]
Consistency limits and number	[Y2]
Grain-size distribution	[Y3]
Grain density	[Y4]
Grain shape and roughness of grain surface	[Y5]
Organic content (ignition loss or wet oxidation)	[Y6]
Calcium carbonate content	[Y7]
Density on placement	[Y8]

2.2 Other Wastes

Other wastes should be described in such a manner that their mechanical behaviour can be defined. Some characteristic parameters for waste materials within operating landfills may be determined by reference to Table 3-6.1.

3 Stress Deformation Behaviour

For fine-grained, soil-like waste mechanical properties (e.g. time dependant compressibility, shrinkage and swelling behaviour and shear strength) are determined in accordance with R 3-1, Table 3-1.2, for the purpose of assessing deformation behaviour and for stability analyses.

For mixed and coarse-grained soil-like waste, triaxial compression tests on large samples may be carried out if necessary. As a first approach, shear parameters and calculated densities may be taken from the tables given in [Z], taking account of the material classification in accordance with [N].

For other wastes, it is necessary to undertake large-scale tests on trial areas, accompanied by stability tests in accordance with R 2-6. Tests using large-scale shear box apparatus or triaxial equipment may also be useful. In this respect the extent to which bio-chemical changes in the waste influence stress deformation behaviour must be examined.

4 Determination of Placement Criteria

Placement criteria, as defined by the qualified geotechnical expert, should be based on investigations to define classification and stress deformation behaviour.

Where soil-like and other wastes are deposited on the same site, the geotechnical expert should specify their sequence of placement and relative quantities.

For fine grained and mixed-grained soil-like waste, the minimum consistency required or the maximum permissible water content must also be specified in relation to stress deformation and shear behaviour.

Placement criteria should be checked under site conditions by means of trial tipping and compaction and adapted, if necessary, to geotechnical requirements in agreement with the site operator.

R 5 Recommendations on Quality Assurance

R 5-1 Quality Assurance Principles

1 General

To ensure the quality of the overall structure of a landfill, the individual components must meet the quality standards. Quality assurance must relate to both the quality of the materials used and to the quality of the workmanship in accordance with the existing state of technology.

2 Quality Assurance Plan

As part of the technical design, the scope of the quality assurance plan and details of checks must be drawn up in accordance with R 2-1.

Requirements concerning the materials and construction methods to be used must also be laid down in the quality assurance plan. The results of the quality assurance tests form part of the checking and commissioning procedure.

Evidence of suitability should be produced for all construction materials and methods, for instance in the case of basal seals or surface capping the suitability tests set out in R 3-1 or R 3-3 are to be conducted. Before construction begins, the suitability of the sealing materials, equipment and the methods to be used should be tested under field conditions, for instance in a field trial in accordance with R 3-5. The results of the suitability tests should be adopted as reference values in the quality assurance plan for the construction work.

The following should be specified in this plan for construction work:
- responsibilities and tasks of the construction supervisors;
- description of the lining system stating the processes to be inspected;
- type and number of quality tests to be undertaken on the construction materials supplied (initial test), on their processing (processing test) and on the completed component (commissioning test).

3 In-house Testing, External Testing and Control Supervision

Quality assurance should comprise [AA]:
- in-house testing by the contractor [AA1];
- external testing by an independent party [AA2].

If appropriate, the regulatory authority may request tests on random samples. All tests must be supervised by a geotechnically qualified expert with extensive knowledge in the field of waste disposal techniques.

These tests comprise:
- initial testing of the construction materials to be processed;
- tests on the processing of the materials;

- supervision of all work, material characteristics and functions which determine quality.

The method of testing adopted, presentation of test data and in-house/external review should be adopted to meet the requirements of the particular construction process.

Acceptance should be based on the results of external testing. Where sections of work are accepted it is necessary to ensure that these are not adversely affected by subsequent construction work. Full-time site supervision by the contractor's inspector and by an external inspector is required. All testing, both in-house and external, should be fully documented.

The holder of the planning permission for a specific waste disposal site should apply to the regulatory authority for final acceptance. This should be supported by full documentation of results relating to the site, and include tests relating to:
- construction of elements of the work and the completed structure;
- adherence to the requirements of the quality plan.

R 5-2 Quality Assurance of Subgrade, Mineral Capping and Basal Sealing Layers

1   General

The sealing system for a landfill comprises all the sealing elements from the base up to the surface of the tip. A sealing system comprises several of the following elements in differing combinations:
- subgrade;
- sealing layers;
- protective layers;
- drainage system;
- gas-venting system;
- if necessary, transitional layer.

Each element should be subject to quality assurance in accordance with the general principles set out in R 5-1.

2   Subgrade

The subgrade represents the base which supports the sealing system. The suitability and settlement behaviour of the in situ soil, and of the waste in the case of the capping seal, should be checked and considered in the design. If necessary, investigations should be specified in connection with quality assurance once the site has been stripped or checked on the surface of the waste once the site has been filled.

The following should be demonstrated by means of in-house and external testing:
- quality characteristics of the subsoil as a stabilising element for the site (R 1-1 or R 2-1) in accordance with the site licence or permit;
- adequate bearing capacity of the subsoil surface and the waste body surface;
- adherence to allowable tolerances in respect to evenness of the subgrade and adherence to design levels and dimensions.

3 Seal

3.1 Scope of Testing

After acceptance of the subgrade and successful completion of the suitability test in accordance with R 3-1, together with investigations on the field trial in accordance with R 3-5, work can commence on construction of the mineral sealing layer. To ensure quality is achieved, field and laboratory tests should be conducted.

The following range of tests will normally be required. In certain circumstances other tests, if proven, may be acceptable:
- characteristics of the materials to be used, including grain-size distribution, consistency limits, water intake and moisture content (every 1000 square metres);
- moisture content on placement, homogeneity of the material placed, number of passes with the roller, quantity of water added, if any (every 1000 square metres);
- minimum clod size, cutting depth and quantity of additives or dosage in the case of multiple component mixtures (every 1000 square metres);
- thickness of the individual lifts, evenness of the lift surfaces and adherence to proposed levels and dimensions (every 500 square metres);
- degree of compaction and homogeneity achieved in the sealing layer for each lift by determination of density, moisture content, grain-size distribution and plasticity, if appropriate, and by survey (every 1000 square metres);
- determination of the permeability of the sealing layer for each lift (every 2000 square metres).

In order to achieve tighter control over the sealing layer, it may be necessary – particularly in the case of non-uniform construction materials – to reduce the size of the test grid. More extensive investigations may be specified in particular cases, depending on the type of seal.

3.2 Notes on Testing

Random samples of the selected materials should be taken at source, on delivery to site and again on placement. The test results should be compared with the suitability test data. In addition, for fine-grained soil material, it is necessary to ascertain whether the soil supplied is sufficiently uniform. Uniformity must relate both to the composition and consistency of the soil. The thickness of each lift should be checked before and after compaction.

The clod size of cohesive sealing materials should be checked in order to achieve an homogeneous sealing layer. Compliance with specified density and moisture content

R 5-2 Mineral Sealing Layers

values is not, by itself, sufficient. Where pressure-mixed sealing materials of mixed grain size are used, it is necessary to check whether the homogeneous mixture achieved is maintained on transport, placement and compaction.

In cases where additives are used, mixed either with the in situ soil or with an imported soil, the quantity and even distribution of the additives should also be checked by means of a grid system. Spacing should be determined for individual tests by reference to the spreading equipment used. During compaction of such materials the cutting depth and homogeneity of the mix should be checked. The cutting depth should be sufficient (a minimum of 3 cm) to ensure bonding into the upper zone of the underlying lift. When construction commences, a relationship should be established between compacted lift thickness and cutting depth on the basis of a field trial were varying moisture contents can occur, this measurement should be related to the range of moisture content anticipated. Adherence to the required moisture content, together with any measures required to achieve it, should be checked.

Each sealing lift should be properly compacted. During compaction the number of passes with the roller, by reference to data from the field trial, and the uniformity of compaction should be checked. To determine the degree of compaction non-destructive testing should be undertaken. Compaction tests should be undertaken on all lifts of the sealing layer. This will allow a qualitative statement to be made regarding the interface between individual lifts. For a sealing layer of fine-grained material, the degree of compaction should be determined from undisturbed samples. Other reliable methods may be used, if they are appropriately calibrated. For mixed grain-size soils an alternative method may be used. If a nuclear probe is used, where the results may be influenced by the type of mineral, the instrument should be calibrated by comparison with data from a sufficient number of density tests (by means of sand replacement or undisturbed sampling).

Moisture content should be checked during construction. It should be determined by oven drying, and in some cases by using a microwave. The nuclear probe can also be used if calibrated regularly.

Permeability should be determined either in the laboratory in accordance with R 3-1, section 5, or in the field provided an appropriate technique is available. The permeability results should be compared with target values related to the test method. When recompacting samples of mixed grain-size soils for permeability testing, the moisture content and density should be selected to correspond to field test data. A method should be found which prevents unacceptable delay to the construction work. The permeability test may not be required prior to acceptance provided other test results relating to quality assurance – particularly grain-size distribution, moisture content and dry density – correspond to data from the suitability tests (R 3-1). Permeability tests are then undertaken for record purposes only.

If weaknesses in the seal are identified from the measuring grid, additional measuring locations should be specified in order to delimit and improve the lower quality zones. A method of backfilling and sealing all sampling locations in the completed liner must be agreed in advance of the construction work to ensure the integrity of the seal.

R 5-3 Quality Assurance for Vertical Cut-off Walls

1 General

Vertical cut-off walls comprise:
- one-phase diaphragm walls;
- two-phase diaphragm walls;
- thin walls.

The following requirements relate to in-house and external testing (R 5-1).

The stated minimum scope of testing applies to in-house testing. External testing should be undertaken by random sampling.

The minimum scope of triaxial compression and permeability testing required for commissioning is 4 tests for each project or:
- 1 test per 3,000 square metres of wall, or
- 1 test per 14 days of wall construction.

2 One-Phase Diaphragm Walls

2.1 Initial Testing

All construction materials such as water, bentonite, mineral fillers, hydraulic binding agents, etc., should be checked on delivery to the site. The characteristic values found should be compared either with those for the construction materials on which the design was based, or those which were used in a suitability test in accordance with R 3-2. Reference samples should be taken from the materials supplied. Additional tests are set out in Table 5-3.1.

2.2 Construction Testing

The prepared compound should be regularly checked. The characteristics obtained should be compared with those of the compound on which the design was based, or which was used in the suitability test. The scope of testing should be as set out in Table 5-3.2 on page 50.

Note: one-phase and two-phase diaphragm walls are a succession of primary and secondary *panels*. These panels can themselves be executed:
- by a single *excavation unit*;
- or by (as is common practice) *three excavation units* (two primary and one secondary). See Fig. 5-3.1 on page 51.

Special care needs to be given to the sampling programme when sealing highly contaminated soils and/or highly heterogeneous subsoil conditions. In these cases, it can be appropriate to execute the laboratory testing programme for each excavation unit and not for each panel.

R 5-3 Vertical Cut-off Walls

Table 5-3.1
Initial tests on construction materials for cut-off walls

Material	Test	Test equipment	Frequency
Drinking water	N/A	N/A	N/A
Industrial water	pH-value conductivity overall hardness	pH paper pH-meter conductivity meter chem. analysis	1 x before building begins, then repeated regularly
Bentonite	Acc. to [U] – yield value – filtrate water loss water intake capacity	pendulum equipt. viscosity meter filter press *Enslin-Neff*	1 x per batch delivered + reference samples
Mineral filler	water intake capacity proportion of grains over 0.125 mm diameter, if appropriate	*Enslin-Neff*, alternative: filtrate water yield, precipitation behaviour sieve	1 x per batch delivered + reference samples
Hydraulic binding agent	Blaine value and proportion of slag on delivery note	if necessary, have the values checked by a cement laboratory using reference samples	
Ready-mixed	Acc. to [U] – yield value – filtrate water loss – run-out time – density water intake capacity	pendulum equipt. viscosity meter filter press Marsh funnel mud balance pycnometer *Enslin-Neff*	1 per batch delivered + reference samples

Table 5-3.2
Tests during construction of one-phase cut-off walls

Material	Test	Test equipment	Frequency and sampling point
Sealing compound	Acc. to [U] - yield value - filtrate water loss - run-out time - density - sand content (not a inflow)	pendulum equipt. viscosity meter filter press Marsh funnel mud balance sand content meter acc. to API	at the inflow 2 x per panel, min. 3 x per shift from the wall cavity, from top and bottom of wall per 250 m² of wall area (125 m² per test)
	permeability ratio uniaxial compressive strength	acc. to R 3-2	from the wall cavity, from top and bottom 1 x per 1000 m² wall area (500 m² per test)
	Note: To obtain the samples a sampling device must be used which ensures that the material is taken from the desired depth of the wall element.			
Accuracy of position	sequence of depth of wall degree of embedment of foot of wall verticality of wall	excavation plumbing excavation, sampling plumbing with 2 lines on grab blades or use inclinometer	continuous 1 x per excavation unit 1 x per excavation unit 1 x per excavation unit
	degree of overlap of wall elements	plumbing with 2 lines on grab inclinometer	1 x per panel
	Note: In the case of continuous excavation with deep diggers, check once every 10 m of wall length			

R 5-3 Vertical Cut-off Walls 51

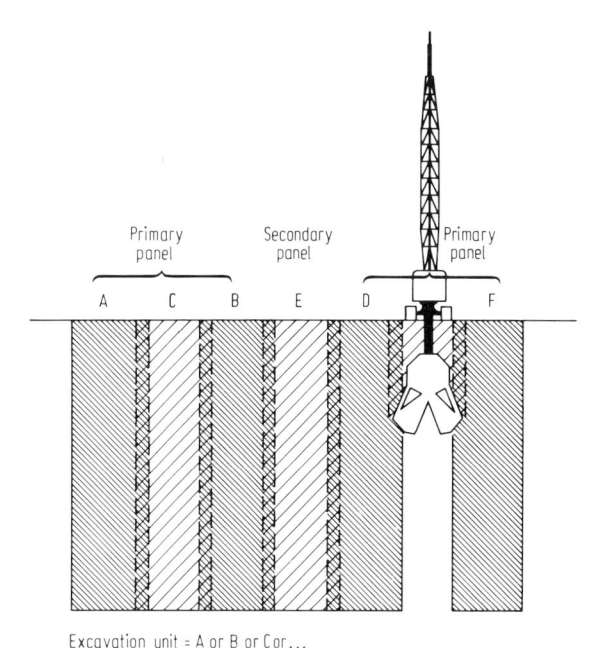

Fig. 5-3.1
Panel and excavation unit

Excavation unit = A or B or C or...

3 Two-Phase Diaphragm Walls

3.1 Initial Testing

All construction materials for phase 1 (suspension) and phase 2 (sealing compounds) should be checked on delivery. For this reason section 2.1 applies mutatis mutandis.

3.2 Construction Testing

The mixes for phase 1 (suspension) and phase 2 (sealing compound) should be regularly checked. The results should be compared with those in the design or suitability test. The scope of testing should be as set out in Table 5-3.3.

Should other construction materials be used, deviations from Table 5-3.3 may be appropriate. For instance, when grouts or slurries are used as a sealing compound in phase 2, the measurement of unit mass can be carried out with a mud balance. The concept of slump can be replaced by a combination of Marsh funnel, run-out time and yield value. It is advisable to introduce the concept of differential unit mass and viscosity between phase 1 and phase 2 compounds.

Table 5-3.3
Tests during construction of two-phase plastic concrete cut-off walls

Material	Test	Test equipment	Frequency and sampling point
Phase 1 (water/ bentonite)	Acc. to [U] - yield value - filtrate water loss - density acc. to [U] - yield value - density	pendulum equipt. viscosity meter filter press mud balance pendulum equipt. viscosity meter mud balance	at the inflow 1 x per shift from the cavity in the sealing wall 1 x per panel approx. 0.3 m above base of cavity before substitution for phase 2
	Note: To ensure total substitution, the yield value must not exceed 70 Pa and the density 13 kN/m³ [BB].		
Phase 2 (sealing compound)	- density - slump factor	sample cube mould (15x15x15cm) and scale flow table	from delivery 1 x per 250 m² of wall area
	Note: To ensure total substitution the density must not be less than 18 kN/m³.		
	permeability ratio uniaxial compressive strength	acc. to R 3-2	from delivery 1 x per 1000 m² of wall area
Accuracy of position	sequence of bedding depth of sealing wall degree of embedment of wall verticality of sealing wall	excavation plumbing excavation, sampling plumbing with 2 lines on digger blades or use inclinometer	continuous 1 x per excavation unit 1 x per excavation unit 1 x per excavation unit
	degree of overlap of wall elements	plumbing with 2 lines on digger blades or use inclinometer	1 x per panel
	Note: These measurements must be taken before the sealing compound is placed.		

R 5-3 Vertical Cut-off Walls

4 Thin Walls

4.1 Initial Testing

The initial testing on all wall construction materials should be executed in accordance with section 2.1.

4.2 Construction Testing

The thin wall sealing compound should be regularly checked. The figures obtained should be compared with those from the design or suitability test. The scope of testing should be as set out in Table 5-3.4.

Table 5-3.4
Tests during construction of thin walls

Material	Test	Test equipment	Frequency and sampling point
Thin wall	– density – permeability ratio – uniaxial compressive strength	mud balance acc. to R 3-2	at inflow 3 x every shift at inflow 1 x ever 1000 m^2 of wall area
Accuracy of position	reaching of bonding level verticality	path-time diagram of vibrating plank pressure/quantity diagram of injection pump inclinometer on the hanging-leader check with 2 m spirit level on vibrating plant	continuous continuous
Survey measurement on sealing compound		pressure/quantity graph with continuous recording	continuous

To confirm the degree of overlap and completeness of all the vibration units the top of the completed thin wall should be exposed (see Fig. 5-3.2).

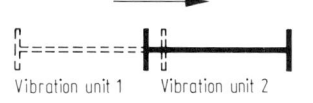

Fig. 5-3.2
Overlap of vibration units

5 System Verification

System verification on the cut-off walls should be achieved by pumping tests. For completed single or double cut-off walls and cut-off wall boxes the test should be carried out by lowering the water table within the isolated area. In the case of sections of single cut-off walls the groundwater level should be lowered, if possible, down gradient.

R 5-4 Quality Assurance for Placement of Waste in External Stability Zones of Landfills

1 General

In order to guarantee quality in the placement of waste within the external stability zones of landfills, (i.e. in the slope zone and the surface zone) it is necessary to set limit values in accordance with R 2-6 and to conduct a suitability test in accordance with R 3-6 specifying the placement criteria. A distinction should be made between soil-like and other wastes.

2 Identity Check

The identity check ensures that only suitable wastes are deposited on the site. In addition to chemical checks it is recommended that the following tests are carried out:
- visual assessment in accordance with the descriptive characteristics laid down in R 3-6;
- in the case of fine-grained soil-like waste, additional consistency check using a simplified field test [CC] and water content determination;
- comparison of details given on the delivery note with suitability test data;
- if necessary, reference samples for water content and consistency testing in the laboratory.

3 Classification

The geotechnical classification of a waste delivered to a landfill provides a basis for defining how and where it should be placed to achieve stability. Special attention should be given to the external stability zones, particularly if wastes of varying suitability are delivered; if necessary, geotechnical classification should be used as a basis for rejecting the material.

R 5-4 Placement of Waste

Classification criteria, in relation to the identity check, include:
- the composition of the waste in accordance with the placement criteria laid down in R 3-6;
- in the case of fine and mixed-grained soil-like wastes, the consistency according to the simplified field test [CC].

A quick decision on geotechnical classification can be made by using prepared check lists.

4 Scope of Testing

The frequency and scope of identity checking should be defined by the qualified geotechnical engineer. It may be necessary in certain cases for a complete identity check to be carried out on each delivery of waste.

Appendix

Geomembranes for Composite Cut-Off Diaphragm Walls

1   General

In some situations (e.g. high concentrations of toxic pollution, significant gas migration into the subsoil etc.) cut-off diaphragm walls with inserted plastic geomembrane sheets are used to reduce permeability and increase chemical resistance and durability of the cut-off barrier.

The geomembrane sheets are normally connected by an interlocking joint system. There are other types of jointing system used such as welding and expansive material elements inserted into the connection.

The toe of the sheets may be embedded in backfill material or in a special grout or concrete.

Two basic methods are used for the installation of geomembranes into the fresh slurry within the diaphragm:
- by stretching the plastic sheets on a heavy installation frame, sinking the frame, disconnecting the plastic sheets and then removing the frame leaving the sheets in place;
- by pulling the plastic sheet in by means of a counterweight on the bottom of the plastic sheet.

Particular care must be given to the design and construction phases of composite diaphragm walls to avoid problems during geomembrane installation, particularly the connection of sheet joints between slurry setting stages and subsequent construction phases.

Special construction procedures should be designed for difficult excavation conditions to prevent curing of the cement bentonite slurry before geomembrane placement.

In certain cases preliminary field trials are suggested.

Geomembrane placement decreases the overall in situ permeability of a standard cement-bentonite slurry wall by about 2 orders of magnitude. The overall permeability can decrease by up to 4-5 orders of magnitude if high quality jointing can be achieved.

2   Suitability Test

The principal geomembrane material used for composite slurry walls is high density polyethylene (HDPE).

As far as tests and controls on the composition of geomembranes are concerned, reference can be made to the sections in this document on geomembranes for landfill lining systems (R 2-3 and R 2-4).

Special care should be given to the jointing elements between sheets. In particular, specific laboratory tests must be undertaken to control the following aspects:
- welds made in the factory, between the joint element (strip) and standard sheet;
- volume increase, when wetted with water and with the leachate to be contained, of expansive plastic elements, (if adopted), to be inserted into the joints during installation to improve the seal;
- stress-strain behaviour of expansive or other types of plastic elements (if adopted) before and after wetting with water and with the leachate to be contained;
- durability of joint elements and chemical compatibility with the leachate to be contained;
- permeability of the joint system to water and to the leachate to be contained.

A detailed report on the composition and characteristics of the geomembrane, together with manufacturers references, must be available prior to the commencement of the works. This should also include test results for material acceptance (see also the sections in this document on geomembranes for landfill lining system, R 2-3 and R 2-4).

3   Quality Assurance

Transport and storage procedures must prevent any damage to the geomembrane sheets due to ultraviolet or other kind of radiation, rupture, puncture, overstress, etc.

All sheets must be numbered and checked prior to installation and the position of each sheet should be recorded in the final, "as constructed" documents.

The minimum scope of testing outlined below applies to in-house testing. External testing should be undertaken by random sampling.

A minimum of 4 geomembrane specimens for every project or 1 specimen for every 10 sheets installed should be provided for stress-strain testing on the basic material and on standard welds made in the factory.

A minimum of 4 specimens for every project or 1 specimen for every 100 metres of joint should be provided for stress-strain and volume-increase tests using in situ available leachate for wetting the specimens.

In situ welds, repairs or other local modifications to joints or geomembrane sheets made on site prior to geomembrane installation must be tested in accordance with instructions from the geotechnical expert.

The location of repairs and/or modifications, together with the location of specimens taken for tests, should be recorded in the final, "as constructed" documents.

The installation procedure should prevent any damage to the geomembrane sheet. Qualified personnel must be selected for this operation.

The diaphragm trench bottom should be carefully checked to ensure it is clean prior to the installation of the geomembrane sheet within each panel.

After installation of the geomembrane sheet the following must be checked:

- the geomembrane must be kept more or less in the centre of the trench and without significant wrinkles by using a suspension system until the slurry has cured;
- the toe of the geomembrane must be kept in close contact with the bottom of the trench using an appropriate counterweight if necessary.

For every project or every 3000 m² installed, a sample of the joint between sheets should be taken at a minimum depth of 3 m (by excavation and after the cement-bentonite slurry has cured) to confirm the quality and location of all the joint elements.

This check is particularly important for long-term joints and for in situ hot welded joints.

In some cases, checks with special cameras can be undertaken if the specific joint system allows.

References

General

[1] Kanitz, J.: Erkundung von Schadstoffausbreitung durch Bodenluftuntersuchungen. Seminar über Altlasten und kontaminierte Standorte. Erkundung und Sanierung. Ruhr-Universität 1985, S. 105–116.

[2] Hueber, D.: Erfahrungen mit Bodenluftmessungen bei CKW-Schadensfällen. Wasser und Boden 1985, H. 5, S. 233–238.

[3] Richtlinie über Deponiebasisabdichtungen aus Dichtungsbahnen. Landesamt für Wasser und Abfall NRW, Düsseldorf 1985.

[4] Burmeier, H.: Arbeiten im Bereich kontaminierter Standorte – Maßnahmen zum Schutz der Beschäftigten. Die Tiefbau-Berufsgenossenschaft 1987, H. 9.

[5] Drescher, J. (1990): Standsicherheit von Deponiekörpern, Festigkeit des Deponiegutes. Vortrag beim 6. Nürnberger Deponieseminar.

[6] Neff, H. K.: Großdeponie Dreieich-Buchschlag. Baugrundtagung Düsseldorf 1984, S. 205–277.

[7] API American Petroleum Institute 1974: Standard Procedure for Testing Drilling Fluids API RP 13B, 5th Ed.

[8] Mehlich, A.: Determination of cation- and anion exchange properties of soils. Soil Sci. 66 (1948), S. 429–445.

[9] Tennant, C. B. and Berger, R. W.: X-ray determination of dolomitecalcite ratio of carbonate rocks. Amer. Mineralogist 42 (1957), S. 23–29.

[10] Pichler, E.: The expansion of soils due to the presence of clay minerals as determined by the adsorption test. Proc. 3 Intern. Conf. Soil Mech. Found. Eng. 1 (1953), S. 43.

[11] ATV-Merkblatt Nr. A 301. Klärschlammeinbau in Deponien. 1988.

Note:
The abbreviation "n/a" on the following pages means "not available".

France

[A] 1. Laboratoire Central des Ponts et Chaussées, Reconnaissance géologique et géotechnique des tracés des routes et autoroutes. Note d'information technique, 1982.
2. Service d'Etude Techniques des Routes et Autoroutes. Laboratoire Central des Ponts et Chaussées, Recommandations pour les terrassements routiers (3 fascicules), Janv. 1976.
3. Magnan, J.-P.: Classification géotechnique des sols 1. A propos de la classification LPC, Bull. Liaison Labo. P. et Ch. - 105 - Janv.-Fév. 1980, pp. 49-52.
4. NF P 94-010: Géotechnique. Glossaire.

[B] 1. Laboratoire Central des Ponts et Chaussées, Reconnaissance géologique et géotechnique des tracés des routes et autoroutes. Note d'information technique, 1982.
2. Laboratoire Central Technique des Routes et Autoroutes. Laboratoire Central des Ponts et Chaussées, Recommandations pour les terrassements routiers (3 fascicules), janv. 1976.
3. Magnan, J.-P.: Classification géotechnique des sols 1. A propos de la classification LPC, Bull. Liaison Labo. P. et Ch. - 105 - Janv.-Fév. 1980, pp. 49-52.

[C] 1. Laboratoire Central des Ponts et Chaussée, Reconnaissance géologique et géotechnique des tracés des routes et autoroutes, note d'information technique, 1982.
2. Service d'Etudes Techniques des Routes et Autoroutes. Laboratoire Central des Ponts et Chaussées, Recommandations pour les terrassements routiers (3 fascicules), Janv. 1976.
3. Magnan, J.-P.: Classification géotechnique des sols 1. A propos de la classification LPC, Bull. Liaison Labo. P. et Ch. - 105 - Janv.-Fév. 1980, pp. 49-52.
4. NF P 94-113: Essai de pénétration statique.
5. NF P 94-114: Essai de pénétration dynamique type A.
6. NF P 94-115: Sondage au pénétromètre dynamique type B.
7. NF P 94-116: Essai de pénétration au carottier.
8. NF P 94-121: Essai CBR en place.
9. NF P 94-110: Essai pressiométrique Ménard.
10. NF P 94-111: Essai au presiomètre autoforeur.
11. NF P 94-112: Essai scissométrique en place.
12. NF P 94-120: Essai au phicomètre.

[D] 1. Laboratoire Central des Ponts et Chaussées, Reconnaissance géologique et géotechnique des tracés des routes et autoroutes, note d'information technique, 1982.
2. Laboratoire Central Technique des Routes et Autoroutes. Laboratoire Central des Ponts et Chaussées, Recommandations pour les terrassements routiers (3 fascicules), janv. 1976.
3. Magnan, J.-P.: Classification géotechnique des sols 1. A propos de la classification LPC, Bull. Liaison Labo. P. et Ch. - 105 - Janv.-Fév. 1980, pp. 49-52.

[E] 1. Laboratoire Central des Ponts et Chaussées, Reconnaissance géologique et géotechnique des tracés des routes et autoroutes, note d'information technique, 1982.
2. Laboratoire Central Technique des Routes et Autoroutes. Laboratoire Central des Ponts et Chaussées, Recommandations pour les terrassements routiers (3 fascicules), Janv. 1976.

3. Magnan, J.-P.: Classification géotechnique des sols 1. A propos de la classification LPC, Bull. Liaison Labo. P. et Ch. – 105 – Janv.-Fév. 1980, pp. 49–52.

[F] 1. Laboratoire Central des Ponts et Chaussées, Reconnaissance géologique et géotechnique des tracés des routes et autoroutes, note d'information technique, 1982.

[G] n/a

[H] 1. Laboratoire Central des Ponts et Chaussées, Reconnaissance géologique et géotechnique des tracés des routes et autoroutes, note d'information technique, 1982.

[I] 1. Laboratoire Central des Ponts et Chaussées, Reconnaissance géologique et géotechnique des tracés des routes et autoroutes, note d'information technique, 1982.
2. Laboratoire Central Technique des Routes et Autoroutes. Laboratoire Central des Ponts et Chausseées, Recommandation pour les terrassements routiers (3 fascicules), Janv. 1976.
3. Magnan, J.-P.: Classification géotechnique des sols 1. A propos de la classification LPC, Bull. Liaison Labo. P. et Ch. – 105 – Janv.-Fév. 1980, pp. 49–52.

[J] 1. Comité Francais des Géotextiles et des Géomembranes. Etanchéite par Géomembranes – Recommandations générales, 1991 (in print).
2. NF P 84-500 (to be published in 1991)
3. Soyez, B., Delmas, Ph., Herr, Ch., Berche, J.-C.: Computer evaluation of the stability of composite liners, Proc. of the 4th Int. Conf. on Geotextiles, geomembranes and related products, Vol. 2, 1990, pp. 517–522.

[K] 1. Cartier, G.: Guide pour les études et les confortements de glissements de terrain, Laboratoire Central des Ponts et Chaussées, 1985, 75 p.
2. Berche, J.-C., Cartier, G., Petal 84: Programme d'étude de la stabilité des talus par ruptures circulaires ou non circulaires. Notice d'utilisation – Laboratoire Central des Ponts et Chaussées, 1984, 37 pages et annexes.

[L] 1. Cartier, G., Baldit, R.: Comportement géotechnique des décharges de résidus urbains, Bull. liaison Labo P et Ch. – 128 – Nov.-Déc. 1983, pp. 55–64.

[M] 1. Fahrhat, H., Faure, R.-M.: Modélisation du fluage des pentes pour le suivi de leur évolution, 3 èmes entretiens Jacques Cartier, Comptes rendus du colloque n° 10, Lyon-Grenoble, 5–9 Nov. 1989.

[N] 1. Service d'Etudes Techniques des Routes et Autoroutes – Laboratoire Central des Ponts et Chaussées, Recommandations pour les terrassements routiers (3 fascicules), Janv. 1976.
2. Magnan, J.-P.: Classification géotechnique des sols. 1. A propos de la classification LPC, Bull. liaison Labo. P. et Ch. – 105 – Janv.-Fév. 1980, pp. 49–52.

[O] 1. NF P 94-057-1: Détermination de la granulométrie des sols par tamisage.
2. NF P 94-057-2: Détermination de la granulométrie des sols fins par sédimentation.
3. NF P 94-051-1: Détermination de la limite de liquidation à la coupelle, de la limite de plasticité au rouleau.
4. NF P 94-055 : Détermination de la teneur en matières organiques par dosage chimique.
5. NF P 94-053 : Masse volumique des sols fins
6. NF P 94-054 : Masse volumique des grains
7. NF P 94-062 : Masse volumique des blocs
8. NF X 31-106 : Détermination de la teneur en carbonate de calcium.

9. NF P 94-060 : Essai au bleu de méthylène d'un sol
10. NF P 94-050-1: Détermination de la teneur en eau pondérale par étuvage
11. NF P 94-056 : Masse volumique des sols par gammadensimétrie des échantillons intacts.

[P] 1. NF P 94-093-1: Essai de compactage Proctor: essai Proctor normal
2. NF P 94-093-2: Essai de compactage Proctor: essai Proctor modifié

[Q] 1. Essais oedométriques, Méthode d'essai L.P.C. n° 13, Juillet 1985.
2. Didier, G., Soyez, B., Heritier, B., Parez, L.: Etude à l'oedomètre du gonflement des sols, çompte rendus de la 9ème Conférence Européenne de Méchanique des Sols et de Travaux de Fondations, Dublin 1987, vol. 2, pp. 549–552.
3. NF P 94-074 : Essai à l'appareil triaxial (in preparation).
   NF P 94-070 : Essai de cisaillement direct à la boîte de Casagrande (in preparation).
   NF P 94-071 : Essai de cisaillement alterné (in preparation)
4. NF P 94-077 : Essai de compression uniaxial (in preparation)

[R] 1. NF P 94-092 : Essai de permeabilité sur échantillon triaxial (in preparation)

[S] 1. NF P 94-093-1: Essai de compactage Proctor: essai Proctor normal
2. NF P 94-093-2: Essai de compactage Proctor: essai Proctor modifié

[T] 1. NF P 94-060 : Essai au bleu de méthylène d'un sol

[U] n/a

[V] 1. NF P 15-301 : Définitions, classification et spécifications des ciments.

[W] 1. NF P 18-011 : Bétons, Classification des environnements agressifs

[X] 1. NF P 94-057-1: Détermination de la granulométrie d'un sol par tamisage.
2. NF P 94-057-2: Détermination de la granulométrie des sols par sédimentation
3. NF P 94-051-1: Détermination de la limite de liquidité à la coupelle, de la limite de plasticité au rouleau

[Y] 1. NF P 94-055 : Détermination de la teneur en matières organiques par dosage chimique.
2. NF P 94-053 : Masse volumique des sols fins
3. NF P 94-054 : Masse volumique des grains
4. NF P 94-062 : Masse volumique des blocs
5. NF X 31-106 : Détermination de la teneur en carbonate de calcium
6. NF P 94-056 : Masse volumique des sols par gammadensimétrie des échantillons intacts
7. NF P 94-050-1: Détermination de la teneur en eau pondérale par étuvage
8. NF P 94-057-1: Détermination de la granulométrie d'un sol par tamisage
9. NF P 94-057-2: Détermination de la granulométrie d'un sol par tamisage
10. NF P 94-051-1: Détermination de la limite de liquidité à la coupelle, de la limite de plasticité au rouleau
11. NF P 94-010 : Géotechnique – Glossaire
12. Laboratoire Central des Ponts et Chaussées, Reconnaissance géologique et géotechnique des tracés des routes et Autoroutes. Note d'information technique, 1982.
13. Service d'Etudes Techniques des Routes et Autoroutes. Laboratoire Central des Ponts et Chaussées, Recommandations pour les terrassements routiers (3 fascicules), Janv. 1976.

France

14. Magnan, J.-P.: Classification géotechnique des sols. 1. A propos de la classification LPC, Bull. liaison Labo. P. et Ch. − 105 − Janv.-Fév. 1980, pp. 49–52.

[Z] n/a

[AA] n/a

[BB] 1. Union Technique Interprofessionnelle des Fédérations Nationales du Bâtiment et des Travaux Publics. Guide pour l'étude et la réalisation des soutènements.

[CC] 1. Laboratoire Central des Ponts et Chaussées, Reconnaissance géologique et géotechnique des tracés des routes et autoroutes, note d'information technique, 1982.
2. Laboratoire Central Technique des Routes et Autoroutes. Laboratoire Central des Ponts et Chaussées, Recommandations pour les terrassements routiers (3 fascicules), Janv. 1976.
3. Magnan, J.-P.: Classification géotechnique des sols 1. A propos de la classification LPC, Bull. Liaison Labo. P. et Ch. − 105 − Janv.-Fév. 1980, pp. 49–52.

Germany

[A] 1. DIN 4020: Geotechnische Untersuchungen für bautechnische Zwecke
2. DIN 4021: Baugrund; Erkundung durch Schürfe und Bohrungen sowie Entnahme von Proben
3. DIN 4022: Baugrund und Grundwasser; Benennen und Beschreiben von Boden und Fels

[B] 1. DIN 4021: Baugrund; Erkundung durch Schürfe und Bohrungen sowie Entnahme von Proben
2. DIN 4022: Baugrund und Grundwasser; Benennen und Beschreiben von Boden und Fels

[C] 1. DIN 4022, T3: Baugrund und Grundwasser; Benennen und Beschreiben von Boden und Fels
2. DIN 4021: Baugrund; Erkundung durch Schürfe und Bohrungen sowie Entnahme von Proben
3. DIN 4094: Baugrund; Ramm- und Drucksondiergeräte

[D] 1. DIN 4022, T2: Baugrund und Grundwasser; Benennen und Beschreiben von Boden und Fels; Schichtenverzeichnis für Bohrungen im Fels
2. DIN 4021, T1: Baugrund; Erkundung durch Schürfe und Bohrungen sowie Entnahme von Proben; Aufschlüsse im Boden
3. DIN 4021, T2: Baugrund; Erkundung durch Schürfe und Bohrungen sowie Entnahme von Proben; Aufschlüsse im Fels
4. DIN 4021, T3: Baugrund; Erkundung durch Schürfe und Bohrungen sowie Entnahme von Proben; Aufschluß der Wasserverhältnisse

[E] 1. DIN 4021, T3: Baugrund; Erkundung durch Schürfe und Bohrungen sowie Entnahme von Proben; Aufschluß der Wasserverhältnisse

[F] 1. Merkblatt über Felsgruppenbeschreibung für bautechnische Zwecke im Straßenbau. Köln: Forschungsgesellschaft für das Straßenwesen e. V. (1980)

[G] 1. DIN 4023: Baugrund- und Wasserbohrungen; Zeichnerische Darstellung der Ergebnisse

[H] 1. Richtlinien für die Beschreibung und Beurteilung der Bodenverhältnisse. Bodenerkundung im Straßenbau, Teil 1. Köln: Forschungsgesellschaft für das Straßenwesen e. V. (1968)

[I] 1. DIN 4021: Baugrund; Erkundung durch Schürfe und Bohrungen sowie Entnahme von Proben
2. DIN 4022: Baugrund und Grundwasser; Benennen und Beschreiben von Boden und Fels
3. DIN 4023: Baugrund- und Wasserbohrungen; Zeichnerische Darstellung der Ergebnisse

[J] 1. Richtlinie über Deponiebasisabdichtungen aus Dichtungsbahnen. Landesamt für Wasser und Abfall NRW. Düsseldorf (1985)

[K] 1. DIN 4084: Baugrund; Gelände- und Böschungsbruchberechnungen

[L] 1. DIN 1054 (Neufassung): Beobachtungsmethode

[M] 1. Brauns, J.: Spreizsicherheit von Böschungen auf geneigtem Geländer Bauingenieur 55 (1980), pp. 433–436

Germany

[N] 1. DIN 18 196: Erdbau; Bodenklassifikation für bautechnische Zwecke und Methoden zum Erkennen von Bodengruppen
[O] 1. DIN 18 123: Baugrund; Untersuchung von Bodenproben; Korngrößenverteilung
 2. DIN 18 122: Baugrund; Untersuchung von Bodenproben; Zustandsgrenzen (Konsistenzgrenzen); Bestimmung der Fließ- und Ausrollgrenze
 3. DIN 18 128: Baugrund; Versuche und Versuchsgeräte; Bestimmung des Glühverlusts
 4. DIN 18 124: Baugrund; Untersuchung von Bodenproben; Bestimmung der Korndichte mit dem Kapillarpyknometer
 5. DIN 18 129: Baugrund; Versuche und Versuchsgeräte; Kalkgehaltsbestimmung
 6. Neff, H. K.: Der Wasseraufnahmeversuch in der bodenphysikalischen Prüfung und geotechnische Erfahrungswerte. Bautechnik 65 (1988), H. 5, pp. 153–163
 7. DIN 18 121: Baugrund; Untersuchung von Bodenproben; Wassergehalt; Bestimmung durch Ofentrocknung
 8. DIN 18 125: Baugrund; Untersuchung von Bodenproben; Bestimmung der Dichte des Bodens
[P] 1. DIN 18 127: Baugrund; Untersuchung von Bodenproben; Proctorversuch
[Q] 1. Schultze, Muhs: Bodenuntersuchungen für Ingenieurbauten. 2. Auflage, Berlin/Heidelberg/New York: Springer Verlag 1967
 2. Empfehlung Nr. 11 des Arbeitskreises 19 – Versuchstechnik Fels – der Deutschen Gesellschaft für Erd- und Grundbau. e.V. Quellversuche an Gesteinsproben. Bautechnik 63 (1986), H. 3, pp. 110–114
 3. DIN 18 137: Baugrund; Untersuchung von Bodenproben; Bestimmung der Scherfestigkeit
 4. DIN 18 136: Baugrund; Untersuchung von Bodenproben; Bestimmung der einaxialen Druckfestigkeit
[R] 1. DIN 18 130: Baugrund, Untersuchung von Bodenproben; Bestimmung des Wasserdurchlässigkeitsbeiwertes; Laborversuche
[S] 1. DIN 18 127: Baugrund; Untersuchung von Bodenproben; Proctorversuch
[T] 1. Neff, H. K.: Der Wasseraufnahmeversuch in der bodenphysikalischen Prüfung und geotechnische Erfahrungswerte. Bautechnik 65 (1988), H. 5, pp. 153–163
[U] 1. DIN 4127: Erd- und Grundbau; Schlitzwandtone für stützende Flüssigkeiten; Anforderungen, Prüfverfahren, Lieferung, Güteüberwachung
[V] 1. DIN 1164: Portland-, Eisenportland-, Hochofen- und Traßzement
[W] 1. DIN 4030: Beurteilung betonangreifender Wässer, Böden und Gase
[X] 1. DIN 18 122: Baugrund; Untersuchung von Bodenproben; Zustandsgrenzen (Konsistenzgrenzen); Bestimmung der Fließ- und Ausrollgrenze
 2. DIN 18 123: Baugrund; Untersuchung von Bodenproben; Korngrößenverteilung
[Y] 1. DIN 18 121: Baugrund; Untersuchung von Bodenproben; Wassergehalt; Bestimmung durch Ofentrocknung
 2. DIN 18 122: Baugrund; Untersuchung von Bodenproben; Zustandsgrenzen (Konsistenzgrenzen); Bestimmung der Fließ- und Ausrollgrenze
 3. DIN 18 123: Baugrund; Untersuchung von Bodenproben; Korngrößenverteilung

4. DIN 18 124: Baugrund; Untersuchung von Bodenproben; Bestimmung der Korndichte mit dem Kapillarpyknometer
5. DIN 4022, T1: Baugrund und Grundwasser; Benennen und Beschreiben von Bodenarten und Fels; Schichtenverzeichnis für Untersuchungen und Bohrungen ohne durchgehende Gewinnung von gekernten Proben
6. DIN 18 128: Baugrund; Versuche und Versuchsgeräte; Bestimmung des Glühverlusts
7. DIN 18 129: Baugrund; Versuche und Versuchsgeräte; Kalkgehaltsbestimmung
8. DIN 18 125: Baugrund; Untersuchung von Bodenproben; Bestimmung der Dichte des Bodens

[Z] 1. DIN 1055, T2: Lastannahmen für Bauten; Bodenkenngrößen; Wichte, Reibungswinkel, Kohäsion, Wandreibungswinkel

[AA] 1. DIN 18 200, Abschnitt 3: Überwachung (Güteüberwachung) von Baustoffen, Bauteilen und Bauarten; Eigenüberwachung
2. DIN 18 200, Abschnitt 4: Überwachung (Güteüberwachung) von Baustoffen, Bauteilen und Bauarten; Fremdüberwachung

[BB] 1. DIN 4126, Abschnitt 7.3 und Erläuterungen:
Ortbeton-Schlitzwände; Konstruktion und Ausführung

[CC] 1. DIN 4022, T1: Baugrund und Grundwasser; Benennen und Beschreiben von Bodenarten und Fels; Schichtenverzeichnis für Untersuchungen und Bohrungen ohne durchgehende Gewinnung von gekernten Proben

Italy

[A] 1. A. G. I. (1977): Raccomandazioni sulla programmazione ed esecuzione delle indagini geotecniche.
2. D. M. 27/06/84: Disposizioni per la prima applicazione dell'art. 4 del decreto del Presidente della Repubblica 10/09/1982, n. 915 concernente lo smaltimento dei rifiuti.
3. D. M. 11/03/1988: Norme tecniche reguardanti le indagini sui terreni e sulle rocce, la stabilità dei pendii naturali e delle scarpate, i criteri generali e le prescrizioni per la progettazione, l'esecuzione e il collaudo delle opere di sostegno delle terre e delle opere di fondazione.
4. CIRC. MIN. LL. PP. 24/09/1988, n. 30483: Istruzioni per l'applicazione del D. M. 11/03/1988.

[B] 1. A. G. I. (1977): Raccomandazioni sulla programmazione ed esecuzione delle indagini geotecniche.

[C] 1. A. G. I. (1977): Raccomandazioni sulla programmazione ed esecuzione delle indagini geotecniche.

[D] 1. A. G. I. (1977): Raccomandazioni sulla programmazione ed esecuzione delle indagini geotecniche.

[E] 1. A. G. I. (1977): Raccomandazioni sulla programmazione ed esecuzione delle indagini geotecniche.

[F] 1. I. S. R. M. (1978): Suggested methods for the quantitative description of discontinuities in rock masses. Int. J. Rock Mech. Min. Sci. & Geomech. Abstr., Vol. 15, pp. 319–368.

[G] 1. A. G. I. (1977): Raccomandazioni sulla programmazione ed esecuzione delle indagini geotecniche.

[H] 1. A. G. I. (1990): Raccomandazioni sulle prove geotecniche di laboratorio.

[I] 1. A. G. I. (1977): Raccomandazioni sulla programmazione ed esecuzione delle indagini geotecniche.

[J] 1. U. S.-E. P. A. 600 2-88/052: Lining of waste containment and other impoundment facilities.

[K] 1. U. S.-E. P. A. 600 2-88/052: Lining of waste containment and other impoundment facilities.
2. D. M. 11/03/1988: Norme tecniche riguardanti le indagini sui terreni e sulle rocce, la stabilità dei pendii naturali e delle scarpate, i criteri generali e le prescrizioni per la progettazione, l'esecuzione e il collaudo delle opere di fondazione.
3. CIRC. MIN. LL. PP. 24/09/1988, n. 30483: Istruzioni per l'applicazione del D. M. 11/03/1988.

[L] n/a

[M] n/a

[N] 1. C. N. R. UNI 10006 (1963): Costruzione e manutenzione delle strade. Tecniche d'impiego delle terre.
2. A. G. I (1977): Raccomandazioni sulla programmazione ed esecuzione delle indagini geotecniche.

[O] 1. A. G. I. (1990): Raccomandazioni sulle prove geotecniche di laboratorio.

2. C. N. R. UNI 10014 (1964): Prove sulle terre. Determinazione dei limiti di consistenza (o di Atterberg) di una terra.
3. A. S. T. M. D 2974-84: Standard Test Methods for Moisture, Ash, and Organic Matter of Peat and other Organic Soils.
4. C. N. R. UNI 10013 (1964): Prove sulle terre. Peso specifico dei granuli.
5. A. S. T. M. S 4373-84: Standard Test Method for Calcium Carbonate Content of Soils.
6. n/a
7. C. N. R. UNI 10008 (1964): prove sui materiali stradali. Umidità di una terra.
8. C. N. R. UNI 10010 (1964): Prove sulle terre. Peso specifico reale di una terra.

[P] 1. A. S. T. M. D 698-78: Test Methods for Moisture-Density Relations of Soils and Soil-Aggregate Mixtures Using 5.5-lb (2.49 kg) Rammer and 12-in. (305 mm) Drop.
2. A. S. T. M. D 1557-78: Test Methods for Moisture-Density Relations of Soils and Soil-Aggregate Mixtures Using 10-lb (4.45 kg) Rammer and 18-in. (457 mm) Drop.

[Q] 1. A. G. I (1990): Raccomandazioni sulle prove geotecniche di laboratorio (cap. 2).
2. A. S. T. M. D 4546-85: One-dimensional Swell or Settlement Potential of Cohesive Soils.
3. A. G. I. (1990): Raccomandazioni sulle prove geotecniche di laboratorio (cap. 3, cap. 4).
4. A. S. T. M. D 2166-85: Unconfined compressive strength of cohesive soil.

[R] 1. A. S. T. M. D 2434-68 (1974): Permeability of Granular Soils (Constant Head).

[S] n/a

[T] n/a

[U] n/a

[V] 1. D. M. 3/06/1968: Nuove norme sui requisiti di accettazione e modalità di prova dei cementi.
2. D. M. 31/08/1972: Norme sui requisiti di accettazione e modalità di prova dei leganti idraulici.
3. D. M. 20/11/1984: Modificazione al decreto ministeriale 3/06/1968 recante norme sui requisiti di accettazione e modalità di prova dei cementi.
4. D. M. 9/03/1988: Regolamento del servizio di controllo e certificazione di qualità dei cementi.

[W] n/a

[X] 1. A. G. I. (1990): Raccomandazioni sulle prove geotecniche di laboratorio.
2. C. N. R. UNI 10014 (1964): Prove sulle terre. Determinazione dei limiti di consistenza (o die Atterberg) di una terra.

[Y] 1. C. N. R. UNI 10008 (1964): Prove sui materiali stradali. Umidità di una terra.
2. C. N. R. UNI 10014 (1964): Prove sulle terre. Determinazione dei limiti di consistenza (o di Atterberg) di una terra.
3. A. G. I. (1990): Raccomandazioni sulle prove geotecniche di laboratorio.
4. C. N. R. UNI 10013 (1964): Prove sulle terre. Peso specifico dei granuli.
5. n/a
6. A. S. T. M. D 2974-84: Standard Test Methods for Moisture, Ash, and Organic Matter of Peat and other Organic Soils.

7. A. S. T. M. D 4373-84: Standard Test Method for Calcium Carbonate Content of Soils.
8. A. S. T. M. D 2167-84: Density and Unit Weight of Soil in Place by the Rubber Balloon Method.
   A. S. T. M. D 1556-82: Density of Soil in Place by the Sand-Cone Method.

[Z] n/a

[AA] 1. C. N. R. UNI 8450 (1983): Norme per criteri e prescrizioni e raccomandazioni per un programma di garanzia di qualità. (Impianti nucleari).

[BB] n/a

[CC] n/a

Netherlands

[A] 1. NEN 5773, 577: Site investigation (geological, hydrogeological and geotechnical).
[B] 1. NEN 5773, 5774: Site investigation (including sampling and description of soil).
[C] 1. n/a
[D] 1. n/a
[E] 1. NEN 5766: Groundwater (boreholes, groundwater levels, field tests).
[F] 1. n/a
[G] 1. Individual standards: Presentation investigation results.
[H] 1. Individual standards: Presentation laboratory tests.
[I] 1. Individual standards: Soil sampling techniques.
[J] 1. Draft general administrative order concerning dumping of solid wastes, VROM 1990; Netherlands language. Ontwerp Stortbesluit Bodembescherming; VROM 1990.
2. Direcives for measures for soil protection in connection with storage and dumping activities; VROM, BO-39, 1984 (KRI-TNO). Netherlands language. Richtlijnen ten behoeve van bodembeschermende maatregelen ter zake van opslag- en stortactiviteiten, VROM, BO-39, 1984 (KRI-TNO).
3. Hoeks, J., H. P. Oesterom, D. Boels, J. F. M. Borsten, K. Strijbis, W. ter Hoeven, 1990. Guidelines for design and construction of cappings for waste material dumping and reuse in constructions. Staring Centrum, SC Report 91, Netherlands language. Handleiding voor ontwerp en constructie van eindafdekkingen van afvalen reststofbergingen, Staring Centrum, SC Rapport 91, 1990.
4. Directive for sealing cappings for waste material dumping and reuse in constructions, Draft 1991, Heidemij Adviesbureau. Netherlands language. Richtlijn dichte eindafwerking van afval- en reststofbergingen, Concept 1991, Heidemij Adviesbureau.
5. Der Kinderen, T. A., G. Hamm, J. Molhoek, Draft 1990. Netherlands language. Bouwtechnische aspecten bij het toepassen van geomembranen ter bescherming van het milieu; Concept 1991, KRI-TNO, KIWA.
6. Protocols for the application of plastic lining materials for soil protection; VROM, BO-39, 1984 (KRI-TNO). Netherlands language. Protocollen voor het toepassen BO-39, 1984 (KRI-TNO).
7. Guidance on dumping of wastes; VROM, 1985. Netherlands language. Richtlijn gecontroleerd storten; VROM, 1985.
Quality Requirements for specific plastic lining materials:
KIWA Draft Criteria Nr. 69, Criteria (for quality assessment) for non reinforced HDPE lining materials, 1989. Netherlands language. KIWA Concept Criteria Nr. 69, Criteria voor afdichtingsfolie van PE-HD zonder versterking, 1989.
[K] 1. SC Report 91, 1990: Resistance between individual bedding planes or friction coefficients (capping system), slope stability.
[L] 1. SC Report 91, 1990: Time-induced deformation behaviour of the waste body.
[M] 1. n/a
[N] 1. NEN 5104: Classification materials for mineral seal.
[O] NEN 5104: Soil physical classification of mineral sealing materials for landfills.

Netherlands

1. NEN 2560: Grainsize distribution.
2. SC Report 91, 1990: Consistency limits.
3. NEN 6415/proef 7 Standaard: Organic constituent content.
4. SC Report, 1990: Grain density.
5. NEN 6446, 6470: Calcium carbonate content.
6. Enslin-method: Water intake.
7. NEN 5781, 5782, 5784 (in prep.): Moisture content.
8. NEN 5781 (in prep.): Density (dry bulk density).

[P] 1. Proef 5, Standaard: Density achievable as function of moisture content through Proctor test.

[Q] 1. n/a
2. n/a
3. NEN in preparation (March 1991).
4. available March, 1991.

[R] 1. SC Report 91, 1990: Permeability testing.

[S] 1. n/a

[T] 1. SC Report 91, 1990: Long-term chemical stability mineral seals.

[U] 1. SC Report 91, 1990: Composition and manufacture of bentonite.

[V] 1. n/a

[W] 1. VNG-pakket indicatief bodemonderzoek, NEN 6446, NEN 6470, NEN 5750: Composition of water used in construction.

[X] 1. n/a

[Y] 1. n/a
2. n/a
3. n/a
4. n/a
5. n/a
6. n/a
7. n/a
8. n/a

[Z] 1. n/a

[AA] 1. NEN ISO 9000 series: in-house testing by contractors.
2. NEN ISO 9000 series: external testing.

[BB] 1. n/a

[CC] 1. n/a

Switzerland

[A] 1. Federal Ordinance relating to the treatment of waste (OTW), Dec. 10, 1990.
[B] 1. Federal Ordinance relating to the treatment of waste (OTW), Dec. 10, 1990.
[C] 1. SNV 670 005
     2. SNV 670 008
[D] n/a
[E] n/a
[F] n/a
[G] 1. Federal Ordinance relating to the treatment of waste (OTW), Dec. 10, 1990.
[H] 1. SNV 670 010
[I] 1. SNV 670 800(a).
[J] 1. Federal Ordinance relating to the treatment of waste (OTW), Dec. 10, 1990.
[K] n/a
[L] 1. Federal Ordinance relating to the treatment of waste (OTW), Dec. 10, 1990.
[M] n/a
[N] 1. Federal Ordinance relating to the treatment of waste (OTW), Dec. 10, 1990.
[O] 1. SNV 670 008
     2. SNV 670 335
     3. SNV 670 340
     4. SNV 670 345
     5. SNV 670 808
     6. SNV 670 810(a)
     7. SNV 670 812(a)
     8. SNV 670 814(a)
     9. SNV 670 816
[P] 1. SNV 670 330
[Q] n/a
[R] 1. Federal Ordinance relating to the treatment of waste (OTW), Dec. 10, 1990.
[S] 1. SNV 670 800(a)
[T] n/a
[U] 1. Federal Ordinance relating to the treatment of waste (OTW), Dec. 10, 1990.
[V] n/a
[W] n/a
[X] 1. SNV 670 810(a)
     2. SNV 670 345
[Y] 1. SNV 670 340
     2. SNV 670 345
     3. SNV 670 810(a)
     4. n/a
     5. n/a

Switzerland

    6. n/a
    7. n/a
    8. SNV 670 317, SNV 670 365
[Z] n/a
[AA] 1./2. Federal Ordinance relating to the treatment of waste (OTW), Dec. 10, 1990.
[BB] n/a
[CC] n/a

United Kindom

[A] 1. BS 5930 (1981): 1981, British Standard Code of Practice for Site Investigation
[B] 1. BS 5930
[C] 1. BS 5930
[D] 1. BS 5930
[E] 1. BS 5930
[F] 1. BS 5930
[G] 1. BS 5930
[H] 1. n/a or BS 5930
[I] 1. BS 5930
[J] 1. n/a
[K] 1. n/a
[L] 1. n/a
[M] 1. n/a
[N] 1. BS 1377: 1975 British Standard Methods of tests for soils for Civil Engineering Purposes (please note that this document is being updated and published in 9 parts. Parts 1, 2, 5 and 7 were published during 1990).
[O] 1. BS 1377
[P] 1. BS 1377
[Q] 1. BS 1377
[R] 1. BS 1377
[S] 1. BS 1377
[T] 1. n/a
[U] 1. n/a
[V] 1. n/a
[W] 1. n/a
[X] 1. BS 1377
[Y] 1. BS 1377
[Z] 1. BS 1377
[AA] 1. n/a
[BB] 1. n/a
[CC] 1. BS 1377

Index

abandoned landfill   7, 10
acceptance   47
agitation test   32
aquifers   1, 2
archaeological monuments   3

bailing devices   8
basal lining system   12
basal sealing layers   45
bentonite   30
borehole geophysics   4
boreholes   4

capping system   13, 17, 18
carbonate ratio   35
classification   41
commissioning test   44
compaction   26, 39, 47
composite basal lining system   15
compressibility tests   33
compression test   27
cone penetrometer testing   4
contaminated land   10
contaminated site   7
control supervision   44

deformation behaviour   2, 22, 23
degree of separation   2
design report   12
diaphragm wall   29, 56
diffusion   14
drainage system   16, 17, 19
dye tests   7

earthquakes   1
encrustation   16
erosion   2, 15
excavation unit   48
external stability   22
external testing   17, 19, 44, 45, 48, 57

field tests   46
field trial   37, 38, 39
filter press   32
filter stability   18, 19

filtrate water loss   32
frost   16

gas monitor method   11
gas venting layer   17, 18
geochemical testing   6
geomembrane   14, 15, 17, 18, 29, 56, 57
geophysical procedures   4
geotechnical report   6
geotextiles   19
grain size distribution   28, 34, 40, 42
granular wastes   41
groundwater   5
groundwater abstraction   2
groundwater chemistry   2
groundwater flow   1, 11
groundwater levels   2, 5
groundwater regime   2, 5

hydraulic binding agent   30
hydraulic gradient   27
hydrogeological data   2

identity check   54
in-house testing   17, 19, 44, 48, 57
initial test   44
interface   15
internal stability   22
ion exchange capacity   34, 35

jointing system   56

Kanitz-Selenka procedure   10
karstification   2
Kjehldahl analysis   34

laboratory tests   5, 26, 40, 46, 57
lateral deformation   23
leachate   2, 28
long-term behaviour   28

marsh funnel   32, 51
medical check-ups   20
micro-seismology   4
mine workings   3

mineral basal seals   34, 37
mineral capping   45
mineral fillers   30
mineral sealing compounds   29
mineral sealing layer   14, 15, 18
monitoring borehole construction   8
monitoring locations   7
mud balance   51
multiple piezometers   8

one-phase diaphragm walls   48, 51
overall safety plan   1, 20
overlap   53

panels   51
performance specification   20
permeability   2, 27, 33, 46, 47, 56
permeability coefficient   27
permeability test   27, 28
personnel protection   21
piezometers   5
placement criteria   26, 43
placement of waste   41, 54
principles of design   12
processing test   44
proctor test   26
protective layer   16

quality assurance   44
quality assurance plan   13, 44

regulating layer   17
restoration profile   19
risk assessment   12
root penetration   18
run-out time   32, 51

safety precautions   9
safety provisions   10
sand replacement   47
sealing compound   30, 31, 32, 50, 51

settlement   22, 23
settlement gauge   23
shear strength   22
single piezometers   8
slip resistance   17, 18
soil air sampling   11
soil samples   10
soil-like granular waste   23
soil-like waste   41
soil-physical classification   26
solubility   2, 6
spreading induced stresses   24
stability   22, 41
stability analysis   13, 24
stress deformation behaviour   2, 26, 33, 42
subgrade   45
suitability test   29, 30, 41, 46, 56
suitability testing   25
surface geophysics   4
suspension   51

thin wall   29, 48
transitional layer   16
trial pits   4
trial pitting   4, 5
trial pumping   5
trial tipping   43
triaxial compression test   27
triaxial tests   33
two-phase diaphragm walls   48

vibration units   53

waste body   17, 22, 23
water intake   26
water intake test   30, 35
water sampling pumps   9
weathering   2, 6

yield value   32, 51

The qualified people to talk to about environmental protection.

Philipp Holzmann AG is a leading international construction company and well placed for the engineering, financing, construction and operation of environmental protection projects based on the up-to-date technology.

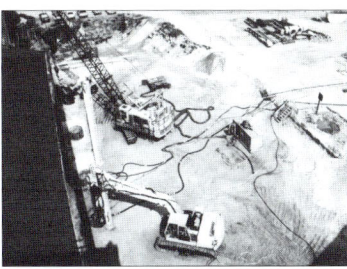

In-situ clining of contaminated soil on a gasworks site

Sealing membrane installed for a household refuse landfill

Shaft-type respository deposition of hazardous waste, constructed for a chemical plant

Holzmann-Harbauer system stationary soil washing plant

PHILIPP HOLZMANN
Aktiengesellschaft

Head office:
Taunusanlage 1 · D 6000 Frankfurt 1
Federal Republic of Germany

TECNIMONT
Gruppo Ferruzzi

CLEAN-UP OF THE ACNA PLANT - CENGIO-ITALY.
1 - Provisional slurry wall drilling
2 - HDPE sheet placing.
3 - Draining trench execution.

Design, construction and operation of industrial facilities & infrastructures.

Worldwide transfer of technologies and know-how in full compliance with environmental regulations and socio-economic constraints.

Environmental engineering, water treatment and supply, waste water, solid and hazardous waste treatment.

Environmental protection, sanitary landfill, monitoring systems.

Engineering and management of built environment and land projects. Environmental impact assessment.

Computer-based integrated information systems for industrial safety management and environmental protection.

Viale Monte Grappa 3 - 20124 Milano - Italy
Tel. 02-6270.1 - Tx 323386 TCM I - Fax 02.6270.9534

Paying the penalty for the sins of the past?

Thanks to our wide-ranging experience in the cleaning of contaminated sites — soil and groundwater — we can ensure a made-to-measure solution to your specific problem, whatever it is.
To remedy the sins of the past we offer the technologies of tomorrow — at today's prices!

Like to know more?
Then let us advice you.

HOCHTIEF
Aktiengesellschaft vorm. Gebr. Helfmann

Rellinghauser Straße 53-57
D-4300 Essen 1 · Germany
Tel. (0201) 824-0 · Fax (0201) 824-2777

Know How!

Contaminated Land:
- risk assessment
- feasibility study and remedial design
- supervision of remediation activities
- research and development

Landfills:
- geotechnical investigations
- design of waste deposits
- gas and water treatment
- safety analysis

Waste Management:
- site selection studies
- environmental impact studies
- waste treatment and recycling
- workers protection

Prof. Dr.-Ing. **Jessberger + Partner** GmbH

Consultants   Bochum · Dortmund · Leipzig · München · Stuttgart

Am Umweltpark 5 · D-4630 Bochum 1 · Tel. (02 34) 6 87 75 - 0 · Fax (02 34) 6 87 75 -10

IG
INGEGNERIA
GEOTECNICA

SGI
STUDIO GEOTECNICO
ITALIANO

SGI/LAB

THE S.G.I. GROUP FOR SOIL AND GROUNDWATER ENVIRONMENTAL STUDIES

Founded in 1964, S.G.I. was rapidly become Italy's topmost private geotechnical consulting firm.

Since then, S.G.I. has diversified its activities in order to be more present in the environmental challenge; the S.G.I. Group is now formed by:

- **S.G.I.** (Studio Geotecnico Italiano), consultants and designers in the fields of geotechnical engineering, engineering geology and hydrogeology, engineering seismology. S.G.I. was present in most of Italy's major designs for landfills and environmental cleanups.

- **I.G.** (Ingegneria Geotecnica), advanced research and consulting center in the fields of soil mechanics and environmental geotechnics.

- **S.G.I. LAB**, the first italian environmental geotechnical laboratory. Investigation on mechanical, hydraulic and chemical behaviour of natural soils, mineral mixtures and solid wastes.

Barricalla, a 600.000 m^3 toxic waste landfill in permeable soil, and Cengio, a 3.5 km long composite cut-off barrier are two recent examples of the S.G.I. group's technical capabilities.

SGI SGI LAB
Via Ripamonti, 89 - 20141 MILANO - tel. 02/5691841 - fax 02/5691845
IG
C.so Montevecchio, 50 - 10129 TORINO - tel. 011/5611811 - fax 011/510568

Erdbaulaboratorium Essen

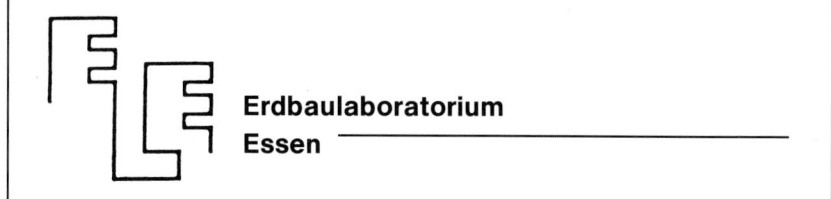

We offer our experiences resulting from more than 40 years of practice in the fields of

- Hydrogeological Studies
- Earthworks
- Recycling
- Waste Disposal and Landfill
- Contaminated Sites and Groundwater

In about 40,000 contracts both home and abroad our engineers and geologists have been involved in

- Investigations on Site
- Laboratory Tests
- Consultation
- Coordination
- Design
- Survey and Checking
- Site Management

Prof. Dr.-Ing. H. Nendza und Partner

Partners: Prof. Dr.-Ing. H. Nendza · Prof. Dr.-Ing. K. R. Ulrichs
4300 Essen 1 · Susannastraße 31
Telefon 0201 / 2 66 08-0 · Telefax 0201 / 25 37 33

SOLETANCHE

Foundations and Underground Construction Specialist

Installation of a HDPE membrane in a cut off wall
(Highway A2 between s'Hertogenbosh and Eindhoven, Netherlands)

6, rue de Watford - F 92000 NANTERRE France - Tel. : 33 (1) 47 76 42 62 - Fax : 33 (1) 47 75 99 10

DYWIDAG presents
Forward-looking solutions for the environmental protection

DYWIDAG-AQUASCHUTZ®-Systems
for keeping the water clean

Separation technology
Sewage treatment
Sewerage technology
and water supply

Systems, concepts and plants
for waste management

Intermediate and final storage
of wastes
Storage of potential water pollution
substances
DYWIDAG-sealing systems

Plants for rehabilitation and
security of waste-loads

DYWINEX-soil filtration plant
DYWIDAG-sealing wall systems

Dyckerhoff & Widmann
Aktiengesellschaft
Erdinger Landstrasse 1
8000 München 81
Phone 0 89 / 92 55 - 1
Fax 0 89 / 92 55 - 21 27

DYWIDAG

BASED ON CONFIDENCE

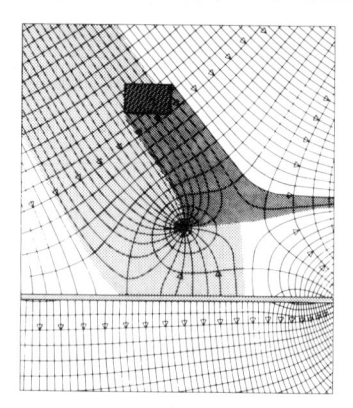

Your choice of consultants is based on their experience, their independence and your confidence. On these principles the Delft Geotechnics Chemistry and Environmental Geotechnology department offers:
- Consulting services and contract research in the field of soil pollution in general and the one resulting from industrial activities in particular.
- The design and evaluation of containment techniques for industrial wastes in landfill situations.
- The design and evaluation of remedial actions in response to soil pollution incidents.
- The evaluation of repositories for medium and low level radioactive waste.

Our consultants are supported a well equipped Field Service and Laboratory. Delft Geotechnics worthy of your confidence!

DELFT GEOTECHNICS
Brassey House, New Zealand Avenue,
Walton on Thames, Surrey , KT12 1QF,
Telephone 0932-253155, Telefax 0932-253658

For waste disposal dumps there are no off the peg solutions.

In 1985, we were the first among geotextile manufacturers to develop needle-punched nonwoven fabrics made of HDPE fibres for the construction of disposal dumps. In the meantime, our Depotex® HDPE protection, separation and filter layers as well as our new Bentofix® sealing mattings have proven their efficiency in innumerable projects. We have learnt a great deal and gathered a wealth of experience. Every waste disposal dump is unique and requires a special, tailor-made solution to guarantee long-term safety. It is exactly this type of solution that we can offer. Take full advantage of our many years of experience, our comprehensive expertise and our high flexibility.

NAUE-FASERTECHNIK

NAUE-FASERTECHNIK
GmbH & Co. KG
Wartturmstraße 1 · P.O. Box 1441
D-4990 Lübbecke 1
Phone: 05741/4008-0
Fax: 05741/400840
Telex: 972178

Microtunnelling

Installation and Renewal of Nonman-Size Supply and Sewage Lines by the Trenchless Construction Method

1989. XIII, 344 pages with 516 figures and 101 tables. 17 x 24 cm. Hardcover DM 198,-   ISBN 3-433-01154-0

The present book is designed to serve as a fundamental source of information on the installation and renewal of nonman-size supply and sewage lines by the trenchless construction method. The descriptions of the presently available nonsteerable and steerable microtunnelling systems, their ranges of application and relevant operational experience gathered, are supplemented by numerous photographs, outline drawings, tables and charts. Knowledge in depth is provided on the subjects of design and operation of different types of soil removal systems, measuring and steering techniques, construction of starting and target pits, as well as obstruction removal.

Moreover, the representations cover the special jacking pipes required for microtunnelling, and furnish detailed information regarding the relevant stress and strain analysis. It is shown how the trenchless construction method can be used for making service house connections, and that there are new ways to link up house service lines with shafts or sewers.

A concluding economic assessment features a comparison of the trenchless construction method with the open trench method.

Beyond that, a comprehensive bibliography is included for reference to special subjects involved with microtunnelling.

Ernst & Sohn
Verlag für Architektur
und technische Wissenschaften
Hohenzollerndamm 170, 1000 Berlin 31
Telefon (030) 86 00 03 19

Today, international competition calls for new endeavour, above all where technological innovations are concerned, inorder to notably increase operating possibilities, performance level and ranges of applications.
"Technology" and "Research" are the company's main strong points, dedicating its efforts in this direction in order to face the challenge of the new century.
One of the main challenges is undoubtedly the safeguarding of the natural heritage.
With a view to this, over recent years, TREVI has been directing efforts towards study, research and applications in the "Environmental sector".

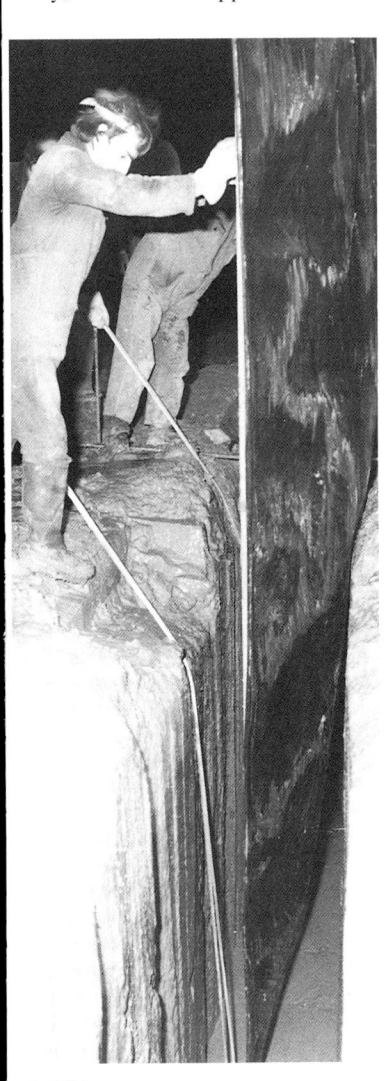

ACNA WORKS IN CENGIO

The ACNA Chimica Organica di Cengio (SV) works, located near the Bormida river, stands on alluvial soil and heterogeneous filling elements over a compact marly sub-stratum which inclines slightly towards the river.
Pollution of the surfacing soils with harmful toxic substances produced through factory activity and through diffusion of the "filtered waste" in the Bormida river, made it necessary to carry out a decontamination operation based on the system of "enclosing" the polluted area.
Encapsulation of the site was performed on the basis of the Technimont project (S.G.I. as Consultants)
using an impermeable vertical barrier socketed into the marly soil formation, integrated with a drainage system capable of keeping down the amount of water within the working area.

The impermeable vertical barriers were made using a cut-off wall in a plastic cement-bentonite mixture, with insertion of HDPE sheet

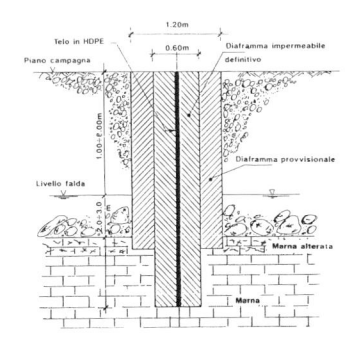

TREVI spa
- **Cesena** - Italy - 5819, Via Dismano - 47023
Tel. 0547/319311-331811 - Telefax 317395
Telex 550687 TREVI I
- **Roma** - Italy - 24, Via G. Nicotera - 00195
Tel. 06/3223068-3217679-3217383
Telefax 06/3217647 - Telex 613345 TECROM I

bonifica

ITALSTAT. GRUPPO IRI

Bonifica, through the use of the most advanced and innovative technologies, supplies services for Engineering Global Systems in the field of territorial physical upgrading and maintenance, intended to cover the entire span of the implementation of projects. Thus, alongside the "classical" services (design studies, supervision of works and technical assistance) in the environmental, hydraulics, transport, agricultural, structures, building construction and economic sectors, the company takes care of procuring funds, defining the legal framework and organizational structures, environmental monitoring, staff training, etc. By this means, it supports the Client in order to attain - with quality - the set target such as for example, the rehabilitation of a historic centre, the economic development of an area, water management, safeguarding a cultural asset, a road, a bridge, a housing estate, and so forth. And to do this Bonifica relies on three critical factors of success which it considers essential and which represent its competitive "advantage": quality of the persons involved, quality of the technologies applied and quality of the organization.

Viale Battista Bardanzellu, 8 - 00155 Roma
Tel. (06)406901 - Telex 620621 Bonger I
Telefax 4063045

ENGINEERING GLOBAL SYSTEMS

Geotextiles in the construction of landfills: To separate, protect, filter, drain and reinforce

nonwovens, wovens, geogrids and composites

For more details, please contact us.

HaTe®

in
landfill construction

HUESKER SYNTHETIC · FABRIKSTRASSE 13–15
P.O. BOX 12 62 · D-4423 GESCHER · GERMANY
TELEPHONE: 49-25 42-7 01-0 · FAX: 49-25 42-7 0137 · TELEX: 8 92 328

Recommendations of the Committee for Waterfront Structures (EAU 1986)

Issued by the Committee for Waterfront Structures of the Society for Harbor Engineering and the German Engineering Society for Soil Mechanics and Foundation Engineering.

5th, English Edition 1986. XXVIII, 521 pages, 195 figures and 43 tables.
14,8 x 21 cm. Hardcover DM 182,-   ISBN 3-433-01065-X

The Committee for Waterfront Structures has worked on an honorary basis since 1949 as a committee of the Society for Harbor Engineering, Hamburg (HTG). Since 1951, the same organization has also operated as Group 7 of the German Society for Soil Mechanics and Foundation Engineering, Essen (DGEG).

Ernst & Sohn
Verlag für Architektur und technische Wissenschaften
Hohenzollerndamm 170, 1000 Berlin 31, Telefon (030) 86 00 03 19